THE NEXT 500 YEARS

THE NEXT 500 YEARS

ENGINEERING LIFE TO REACH NEW WORLDS

CHRISTOPHER E. MASON

THE MIT PRESS   CAMBRIDGE, MASSACHUSETTS   LONDON, ENGLAND

Figures and art by Dr. Matthew MacKay.

This book was set in Stone Serif and Stone Sans by Jen Jackowitz. Printed and bound in the United States of America.

Library of Congress Cataloging-in-Publication Data

Names: Mason, Christopher E., author.
Title: The next 500 years : engineering life to reach new worlds / Christopher E.
    Mason.
Description: Cambridge, Massachusetts : The MIT Press, [2021] | Includes biblio-
    graphical references and index.
Identifiers: LCCN 2020012162 | ISBN 9780262044400 (hardcover)
Subjects: LCSH: Space medicine. | Astronauts—Health and hygiene. | Astronautics—
    Human factors. | Genetic engineering.
Classification: LCC RC1150 .M37 2021 | DDC 612/.0145—dc23
LC record available at https://lccn.loc.gov/2020012162

10   9   8   7   6   5   4   3   2   1

Special Thanks: Dr. Matthew MacKay

Dr. Matthew MacKay not only made the beautiful figures in this book, but he was also cocaptain on the quest to detail the 300-year vision presented here. This vision is a shared hope of what can be done, and what must be done, for our species and all others we serve to guard (past, present, and future). He endlessly helped as editor, writer, debater, and visionary. While the biotechnology, engineering, and genetic guardianship ideas in this book are described in the future, hopeful tense, they are actually grounded in Dr. MacKay's published and pioneering work, which has shown that many of these ideas are in fact possible. Many of the constructs for cells, circuits, and planetary design already have a proof of principle from the writings and algorithms he has published, and this book could not have happened without this guiding light and engine of science.

To all humans and any extinction-aware sentience

CONTENTS

INTRODUCTION: THE EMBRYOGENESIS OF HUMANITY

Embedded in every single neuron in a human brain is a shared ancestry of humans' genetic code—deoxyribonucleic acid (DNA)—carrying the unique capacity for protecting and preserving the complexity and beauty of all life. This DNA also contains the molecular recipe for the synthesis of human bodies, brains, and minds, whose dreams and technologies have spanned visions of other planets and spacecraft that have reached beyond humankind's first solar system. The fundamental thesis of this book is that the same innate, biological capacities of ingenuity and creation that have enabled humans to build rockets to reach other planets will also be needed for designing and engineering the organisms that will sustainably inhabit those planets.

The missions to other planets, as well as ideas for planetary-scale engineering, are a *necessary duty* for humanity and a logical consequence of our unique cognitive and technological capabilities. There is no other species that leverages, or even can leverage, the frailty of mortality into an intergenerational stability of sentience. As far as we know, humans alone possess an awareness of the possibility of our entire species' extinction and of the Earth's finite life span. Thus, we are the only ones who can actively assess the risks of (and prevent) extinction, not only for ourselves but for all other organisms as well. This is unusual. Most duties in life are chosen, yet there is one that is not. "Extinction awareness"—and the need to avoid extinction—is the only duty that is activated the moment it is understood.

This gives us an awesome responsibility, power, and opportunity to become the universe's shepherds and guardians of all life-forms— quite literally a duty to the universe—to preserve life. This means we need to prevent the death of not only our species, but of all species on which we depend and any others we may find that are or were threatened—thus, all current, future, and even past life-forms (through de-extinction). This duty is not only for us, but for any species or entities who can engineer themselves to avoid the end of the universe. Even if our species does not survive, this duty is passed on to the next sentience, which will undoubtedly arise.

Regardless of *who* is here in billions of years (ourselves or someone else), life cannot remain on Earth, because the sun will eventually overheat the Earth, likely engulf the Earth, shrivel into a White Dwarf, and die. Earth is the only home we have ever known, and if it remains that way, it will also be our grave. Thus, it is essential for us to land on, live on, and survive on planets around other stars to continue this duty of humanity. To do this, we will need to deploy all the technological, physical, pharmacological, and medical protective measures that we know and will learn, but we can also, for the first time ever, deploy genetic measures of defense. As a part of this moral duty to preserve and protect life, we will eventually need to engineer it. Evolution has created life only in the context of one planet so far—in the Goldilocks zone of a temperate Earth—and it is likely that we, and all other organisms, will need extensive physical *and* genetic help to survive anywhere else—even if just to arrive at our next destination.

Sending any Earth-evolved organism to any other planet would result in almost certain death, which represents the sad, evolutionary "good luck" plan. This limited plan is not our only option. Today, we know enough to be able to modify, tweak, and engineer life to improve the odds of survival or to create entirely new adaptive features and mechanisms. Evolution has finally created an organism that can direct and engineer not only its own development, but also the evolutionary paths of all other life. This stage of "directed evolution" for life, drawing on all past, current, and future genetic substrates, is an essential step for *life itself* to survive.

To save life, we will need to engineer it. Notably, humans are already *accidentally* engineering life and directing evolution; now it is time to do it with volition, direction, and purpose. Through the use of the collective genetic lessons we have learned from all organisms over billions of years, we have developed many extraordinary technologies that make this possible, and many are highlighted in this book. Our own DNA is composed of relics of what life once was, life as it is today, and the ongoing evolution toward what life will become.

However, with synthetic biology and DNA synthesis costs declining, we can even imagine extinct life returning, as well as means by which to create chimeric or hybrid entities, and this too will be examined in this book. Moreover, by using studies of organisms in extreme environments (extremophiles), we can learn new mechanisms and modalities of adaptation that have enabled alien-like life on Earth, and, indeed, some of this work we have already begun in our laboratory, such as using genes from tardigrades in human cells. These technologies and new methods will enable humans and other organisms to survive in otherwise impossible settings caused by extreme levels of radiation, temperature, or pressure.

This inherent duty of humanity—to preserve life—is as natural as one cell dividing into two. Right now, all humanity is as fragile as an embryo at the single-cell stage. We are an embryo full of extraordinary potential, but only on the primordial beginning step of our home planet. Our next step is to get to a nearby planet (e.g., Mars) and set up a sustainable habitat in order to ensure we have a backup plan for all life, including humanity. This accomplishment would be a point of euphoric celebration, as the tired eyes of a Martian explorer would watch as the sun sets on the dusty horizon, and the air would reveal beautiful blue sunlight diffracting through the thin Martian atmosphere and dust. At long last, we would have two planets to call home around the same sun.

After decades of physical and biotechnological development, we will be able to call many different celestial bodies within our own solar system home. Through this advancement and capability of testing theories across multiple different worlds, we will acquire the ability to launch toward a second sun by 2500. Once we are an interstellar species, we will effectively have a "solar-system backup plan," drastically decreasing the chances of life's extinction. However, this begs inevitable questions: How many stars would we go to? How do we pick? How far will we travel? Indeed, given enough time, fundamental philosophical questions emerge about the endless expansion or inevitable implosion of the universe, and whether or how humanity should alter the structure of the universe as an extension of this duty. These questions will also be addressed in this book (quick preview: yes).

When given the choice between engineering life or facing inevitable death, there is clearly only one path. The right thing to do, in order to survive extinction, is to engineer at a genetic, cellular, planetary, and interstellar scale. This ensures preservation of humanity and, also, of all other life, which may not arise in the next universe or ever again. Our species' unique moral duty is a duty to the universe and to life itself. To protect the universe, we must alter the universe.

To do this, we need a long-term plan. This book will take you through the first 500 years of such a plan, including lessons from bacteria, viruses, and whole planets, as well as from the first astronauts who pushed the limits of human spaceflight.

1

THE FIRST GENETIC ASTRONAUTS

My skin had not touched anything in 340 days. . . . anything it touched, it felt like it was on fire.

Astronaut Scott Kelly

Huddled around glowing monitors full of molecular, genetic, and telemetry data, we were united in our bafflement and concern. We simply could not believe our eyes.

"Are these the highest levels ever seen in a human body?" asked Dr. Cem Meydan. "How did he survive?"

It was a crisp December evening in New York City in 2017, at our genetics laboratory at Weill Cornell Medicine. We had just finished the integrated analysis of all the molecular data (DNA, RNA, proteins, small molecules) from Captain Scott Kelly, who had completed the longest-ever NASA mission in space—almost a complete year (340 consecutive days). Kelly's long-duration spaceflight was part of a unique experiment at NASA called the Twins Study, which leveraged identical twin astronauts (Mark and Scott Kelly) to discern what happens to the human body before, during, and after a year in space. The research spanned ten research teams across the United States; our laboratory worked on the genetic, epigenetic, microbial, and gene-expression analyses. We had comprehensive molecular and genetic data from Scott's time in space, which we could compare to Mark's time on Earth. Our job was to (1) assess what happened to Scott during such a long mission, (2) learn about the changes as a guide for Mars missions, and (3) plan for ways to mitigate future risks to other astronauts.

It was clear that his body did not enjoy the return to gravity. Scott himself described the unpleasantness in his book, *Endurance: A Year in Space, a Lifetime of Discovery*. "My ankles swelled up to the size of basketballs," he noted, amazingly with a calm demeanor. "I felt like I needed to go to the emergency room."

Even though he wanted to go to the emergency room, he knew the reason for the body's changes; he had just returned from space! However, this knowledge did not comfort his immune system. He broke out in rashes all over his body, especially where anything touched his skin. His body was even reacting to something as simple as the weight of clothing being pulled down onto his skin by gravity, causing visible irritation. We could see this immune response in the molecular data from his blood work, especially with changes to his proteins and RNA (gene expression). But we all wondered while staring at the monitors . . . Was this reaction part of a normal readaptation to gravity? Does this have any impact for the plans to go to Mars?

"These are the highest levels of inflammation markers and cytokine stress I've ever seen," I said. "Let's triple-check the data."

We checked with Dr. Scott Smith at NASA, who leads the biochemistry analysis unit for the twins and other astronauts, and he confirmed that the data were correct. He also noted, "This is the highest we've ever seen, by a long shot." Samples were processed in duplicate, just to be sure, and our measurements and computational analyses matched. While inflammation is a normal part of the body's response to stress, here, Captain Kelly's return to gravity catapulted his inflammation markers to unseen heights (figure 1.1).

Specifically, interleukin receptor antagonist 1 (IL-ra1), which is an important natural anti-inflammatory protein, as well as other cytokines, such as IL-6, IL-10, and C-reactive protein (CRP), were all spiking extremely high upon the return to Earth. CCL-2, which is a cytokine (a type of protein that leaves cells to signal other cells) that recruits immune cells to sites of injury or infection, was also spiking very high.

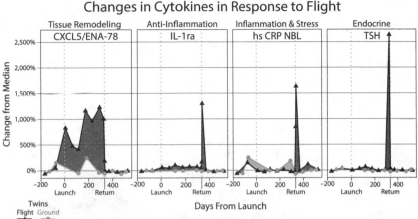

1.1 Many cytokines changed expression during the Twins Study, comparing Scott Kelly's cytokine levels (black) to those of his twin brother, Mark Kelly, who remained on Earth (gray). Dotted lines indicate Scott's launch and return to Earth. Cytokine levels are normalized to their median expression across the analyzed time in both bothers. Some cytokines were elevated throughout the whole mission, such as C-X-C motif chemokine 5 (CXCL5), which plays a role in tissue remodeling. Other molecules primarily spiked upon returning to Earth, such as interleukin-1 receptor antagonist (IL-1ra) and C reactive protein (CRP), which deal with inflammation and thyroid-stimulating hormone (TSH).

We quickly searched across the index of all scientific literature and medical journals to see if anyone had ever seen anything close to these levels, especially for IL-ra1 (>10,000 pg/uL). For IL-ra1, the closest we could find was for patients who had just had a myocardial infarction (a kind of heart attack), from a paper in 2004 (by Patti et al.). For IL-10, spikes were found to be associated with patients who had just survived a severe bacterial infection of the blood (called sepsis).

Somehow, even amid this discomfort, when Scott got back to Earth, he jumped right into his swimming pool and went on to live a normal life in the days and years after. However, these markers were not the only thing that dramatically changed. Other changes could be seen across his tissue systems such as his blood and bones, and we even saw additional molecular changes in his DNA and RNA. We had an unprecedented chance to look at almost everything in the body, from each nucleotide of the genetic code to how cellular responses manifested across Scott's body, resulting in phenotypical changes. Most of these measures were entirely new metrics for any astronaut, including the first complete genetic profiles (genome), as well as other features (figure 1.2) for a spacefaring human. We used all these data to gauge what happened inside the human body during a year in space.

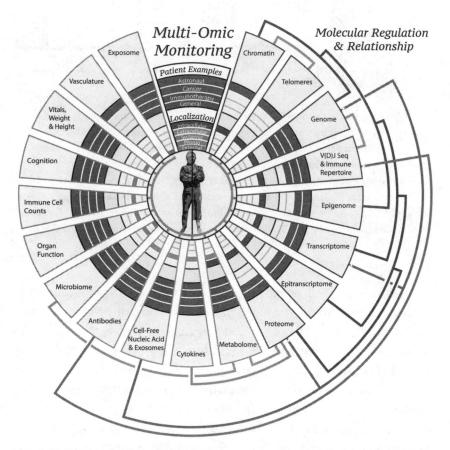

1.2   Multi-omic monitoring platform for astronauts and relation to the clinic: Four monitoring examples are highlighted, including astronauts, cancer patients, immunotherapy patients, and general patients. Each example highlights different -omic data that can be utilized for regular monitoring and follow-up. Molecular interactions between different -omic data demonstrate the need to integrate all these measurements into one platform.

DNA DAMAGE

We first looked at the impact of radiation, which can damage DNA, cells, proteins, and all the regulatory machinery inside cells. Flying at nearly the speed of light are galactic cosmic rays (GCRs), which originate from stars outside our solar system, and solar energetic particles (SEPs), which originate from our sun itself, both sources of radiation that flew through Scott's body. These particles leave a wake of damage like microscopic bullets through the body. GCRs and SEPs are high-energy particles, usually made from protons, helium, and a subset of high-energy ions (HZE ions, which stands for high [H], proton/atom number [Z], and energy [E]). This damage to astronauts was first observed in 1969 and 1970, when Neil Armstrong wore a foil plate around his ankles as he traveled to the moon and back. On this plate, streaks of these HZE particles can be seen displacing the sensor, like marks made by someone drunkenly playing on a high-energy Etch A Sketch or recordings from a nuclear accelerator laboratory after atoms are smashed into each other. Except, in this case, the accelerator is shooting HZE particles, and the laboratory battleground is, unfortunately, the human body.

These HZE particles normally go unnoticed during the day, but they can appear in unexpected places. When Scott closed his eyes to go to sleep at night on the International Space Station (ISS), he could see streaks of light, as if there were shooting stars behind his eyelids. These magical displays of light were actually the HZE particles blasting his retinal cells and passing through his eyes, erupting in a lightshow of beautiful, but terrifying, cellular damage as a bedtime story.

Given such reports, we were all worried about what we would find inside Scott after such a long mission. As it turns out, we had several surprises. One of the first things we expected was that his telomeres would probably break down and shrink from radiation and the stresses that accompany spaceflight. Telomeres are the ends of human chromosomes, which normally shrink as you get older, and their lengths are also associated with both diet and stress. As they disappear, the chromosomes become less stable, contributing to the normal molecular process of aging. Dr. Susan Bailey led the research to test this question, and we sent some of our DNA to her lab, and vice versa, to confirm the results.

UNEXPECTED RESPONSES TO SPACEFLIGHT

Strangely, Scott's telomeres got *longer* when he was in space, which is the opposite of what we expected. We then triple-checked both sample sets of DNA from the Bailey lab and our own lab, and this lengthening was indeed confirmed. It was most pronounced in one type of immune cell called T cells (primarily CD4+ T cells, though evidence was also found in CD8+ T cells), with less evidence of telomere lengthening in B cells (CD19+ cells). Overall, multiple sample replicates, extractions, laboratories, and methods (FISH, PCR, nanopore) confirmed the results, leading us to conclude they were correct.

But then the immediate questions were how and why? We looked at the other data we had collected to make sense of it. Weight loss is associated with telomere maintenance, and Scott did lose about 7 percent of his body weight on the mission because of the rigorous conditions of spaceflight, but he also had daily workouts, nutritionally optimized food, and an absence of alcohol. In some ways, his life in space was healthier than it was on Earth. Also, folic-acid metabolism is linked to telomere maintenance, and the folic-acid levels in Scott's blood were also elevated in flight, adding another possibility. He gained two inches in height during the mission. He also was traveling closer to the speed of light.

Some people got very excited when we first reported these results and asked, "Is space the fountain of youth? Can you get taller and younger if you go to space?" Sort of.

First, we have to isolate all the variables and consider what else happened to him. Scott did travel closer to the speed of light, traveling at an average of 7.68 kilometers per second (km/s), which then enables a calculation using Einstein's relativity and time dilation on a human body. Time dilation occurs when an object moves closer to the speed of light, making time move more slowly for the object in motion relative to the reference frame of other objects. This is dependent on several factors that can be entered into the Einstein/Schwarzschild equation, assuming a few parameters:

(1) A $dr = 0$ (stay at constant radius) and $df = 0$ (same orbital plane);
(2) The ISS orbital speed of 7.68 km/s, with a radius of the ISS at 400 km above the Earth's surface;
(3) The change for Mark Kelly (dt_{MK}) on Earth compared to Scott Kelly (dt_{SK}) on the ISS.

The full equation includes the coordinates of colatitude (theta), the speed of light (c), and the gravitational metric between two spheres (omega), seen here:

$$g = c^2 dr^2 = \left(1 - \frac{r_s}{r}\right) c^2 dt^2 - \left(1 - \frac{r_s}{r}\right)^{-1} dr^2 - r^2 g_\Omega$$

Given this equation, Scott became about 0.1 seconds younger than everyone on Earth, including his brother. Since Scott was born 6 minutes after Mark, this made Scott an additional ~0.1 seconds "younger" than his brother after a year in space. However, even though he is technically younger than what he would have been if he had stayed on Earth, this is not likely a significant factor for his longer telomeres.

We know this because we saw many other modalities of the biology change as well, such as changes in gene expression (off/on or up/down levels of various genes). We all have thousands of genes that change expression every day, so it was not surprising that we could see genes changing when he got to space and when he came back down to Earth. His altered genes' expression included those responsible for DNA repair and cellular respiration. His immune system was also highly activated, including when he received the first-ever flu vaccine in space. Also, we saw evidence of hypercapnia, which is a condition of too much carbon dioxide in the blood and where one can start to feel light-headed and develop a headache; indeed, this irritation was mentioned by Scott in his book. He noted that he got headaches because of the varying carbon dioxide levels, and whenever the CO_2 scrubbers of the space station would break down, he felt as if he had more headaches during these intervals.

We looked at the carbon-dioxide levels on the space station, and though there were some fluctuations, they were not too dramatic and should not have led to physiological changes; we had to look for other causes. As it turns out, breathing in zero gravity is not like breathing on Earth. In particular, every time you breathe out, a small cloud of CO_2 can form in front of your face. This CO_2 minicloud stays by your face, unless you have a fan or move. Thus, some of what we could see in Scott's blood, and likely that of other astronauts, were face-associated, CO_2 miniclouds, more like the atmosphere of Venus than that of Earth.

We also looked at the dynamics in Scott's microbiome, which are the microorganisms (bacteria, viruses, fungi, and other small, nonhuman cells) inside his body. Specifically, we wanted to see what happened to the microbiome during spaceflight. We observed some changes in flight for the ratio of species, specifically for the Firmicutes/Bacteroides (F/B) ratio, using stool data from Drs. Stefan Green, Fred Turek, and Martha Hotz Vitaterna and some of our own data from skin and oral swabs. However, the total diversity was mostly maintained, which is good news. They did eventually return to normal, so there were no big red flags in the microbiome.

But other molecules in Scott's blood did show some unusual features. The mitochondria, which are normally resting inside cells and carrying on cellular respiration to ensure that cells can literally breathe and get energy, were spiking in his blood during the flight—especially when he first got to space. A normal person would have 500 copies of mitochondrial DNA per milliliter (mL) of blood, but Scott showed levels as high as 6,500 copies/mL, based on data from Drs. Kiichi Nakahira and Augustine Choi. We then examined the RNA in the blood, working with Stacy Horner and Nandan Gokhale at Duke University, and there, too, we could see higher levels of mitochondria.

This was an entirely new measure of stress for astronauts, but it has been seen before in other contexts. At Columbia University in New York City, there are laboratories that study extreme variations in mitochondria and even "mitochondrial psychobiology" (in work by Drs. Andrea Baccarelli and Martin Picard), where they have looked in Earth-bound individuals for changes in mtDNA in the blood of people undergoing stressful situations. This includes an interesting study of people who gave speeches in a room full of strangers, where the researchers also observed spikes in the blood's mtDNA levels after the talks. Thus, there is ample evidence that mtDNA can appear after general bodily stress, the anxiety of public speaking, or other senses of danger as well.

But—why would human cells start to produce or eject their own means of energy? Here, too, other studies have given clues as to what was happening during a year in space. A 2018 paper (by Ingelsson et al.) showed that white blood cells (lymphocytes) can eject their mtDNA as a way to prime the immune system. These "DNA webs" serve as a warning sign for other immune cells to prepare to fight an infection or defend against a cellular threat, and it seems these webs work in space just as well as they do on Earth. Work from Afshin Beheshti at NASA and our group has now seen the mtDNA stress appear in multiple astronauts, along with other RNA signatures of spaceflight (including small RNAs called miRNAs). All of these surprises, from telomeres, gene expression changes, hypoxic miniclouds, immune stress, mtDNA, and inflammation, happened quickly and seemed to be a rapid, unexpected response to spaceflight, which hopefully would return to normal.

RETURNING TO EARTH

Fortunately, almost everything is plastic and malleable about the human body's response to long-term spaceflight. While Scott did gain two inches of height, this gain was just from the lack of compression on his spinal column, and his newfound height disappeared within a few hours of returning to Earth. Also, within forty-eight hours, Scott's telomeres had returned to normal length, and most of his blood and physiological markers were within normal ranges. For his gene-expression dynamics, 91 percent of the changes that occurred while he was in flight returned to normal within six months of returning to Earth.

Thus, most of Scott's spaceflight-induced gene expression returned to normal, but not all. Some genes did carry a "molecular echo" of their time in space, still actively working to continue DNA damage repair and maintain DNA stability. These data also matched what we observed when we examined his chromosomes for other breaks or damage. Even after returning to Earth, Scott showed continual signs of low-level inversions and translocations, which are breaks in the chromosomes, that were continually being healed, replaced with newer cells, and genetically fixed.

Even six months later, some genes were still disrupted in their expression—still adapting—and these are the ones we will cover later in the book, when we discuss the long-term plans for human-genome engineering. The gene expression data showed how the body adapts to space and how, sometimes, it does not completely return to normal. This matches what Scott himself mentioned, that he didn't "feel normal" until seven to eight months after being back on Earth. Also, the work from Dr. Matthias Basner showed that Scott's cognitive speed and accuracy were worse after his return to Earth. In our own work at Cornell with David Lyden, we saw proteins that are normally only in the brain appear in the blood, which matched some of the same genes that created those proteins and indicated a change in the blood-brain barrier. Overall, these molecular changes give us a guide as to which genes may need to be accelerated, decelerated, or otherwise altered to help this response to spaceflight.

Other biological features that could also be tweaked come from clues in the cytokine data, specifically the inflammation markers. Some inflammation markers, like IL-6, went up by thousands of percent on the day he landed, and some even higher two days later. The blood work clearly showed a spike of inflammation cytokines that led to so much pain and is likely why Captain Kelly broke out in rashes. These data were also confirmed with cytokine data from Drs. Tejas Mishra and Michael Snyder from Stanford. When we looked all at the markers together as a pathway, the majority of the functions pointed to muscle regeneration. In short, the pain of using his muscles again was forcing a massive restructuring of the body, with his blood printing the molecular receipt of this expensive physiological purchase. In this amazing event of the human body returning to Earth from space, the blood was screaming out, "Oh crap—gravity! I need to use my muscles again!"

Although landing back on Earth was clearly painful, one good thing about Mars is that it has 38 percent of Earth's gravity. Given that difference, the landing might only constitute 38 percent of an "Oh crap!" moment and 38 percent of a challenge to adapt to the surface when landing on Mars. From these results, it seems that a person could actually survive the trip to Mars, and then likely survive the landing, to begin building a new, rust-hued home.

FUTURE MISSIONS

A large caveat of the Twins Study is that we only had two subjects, derived from a single embryo, with only one in space for a longer duration—so we can only extrapolate these results to others in a limited way. Moreover, spending a year on the ISS is still within Earth's magnetosphere, which extends roughly out to 65,000 km, and still acts as a protective shield from radiation for astronauts. To get a sense of the challenge for a mission to Mars, we can compare other missions to the expected amount of radiation astronauts will incur on the way to the red planet, which is about 300 millisieverts (mSv), as well as a 30-month round-trip mission, which is about 1,000 mSv (figure 1.3). This would be more than six times the amount of radiation Scott saw in his mission. While such radiation is not pleasant, there are ways this can be addressed and protected against, which will be revealed in later chapters.

Indeed, we do not have to accept these radiation risks without defending ourselves against them. Though we do already protect astronauts physically, pharmacologically, and medically, these mitigations need to be improved, and we should further use any other means of protection for them as well. Notably, the one biological defense mechanism that has not yet been implemented for astronauts (though it has been for patients on Earth for a wide range of conditions) is genetic engineering.

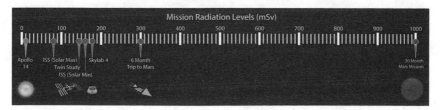

1.3 Radiation metrics for various mission parameters: Estimated and measured radiation metrics for a variety of missions in millisieverts (mSv).

GENETIC DEFENSES

Given the clear risks for long-duration missions to other planets (e.g., Mars) and the challenges of later-stage (e.g., interstellar) missions that would put humans in more dangerous environments with more radiation and less ability to create food and maintain proper metabolism, an exploration into our genetic defenses is warranted. In other words, if we can learn the secrets of all other species and craft a series of genetic protections, we would be embarking on not only a needed means of survival, but also a manifestation of our own genetic duty. We do everything we can to keep astronauts safe through engineering their rockets and ships, but could we make some of the protections on the inside, within the astronauts themselves? Should we do such a thing? Is it right to genetically modify astronauts?

Some of these abstract questions became tangible with He Jiankui, who began to genetically modify human embryos using CRISPR (discussed more in later chapters), two of whom were born in 2018. He did all the work in secret and misled the Institutional Review Board (IRB) at his university, kicking off an angry response when he decided to bring gene-edited babies into the world.

Such a process of bringing groundbreaking medical technologies into the world is the absolute worst way to do it—in secret with little oversight—but the idea is no longer hypothetical. The question now is: How do we actually start to regulate genetically engineering embryos or make sure it doesn't go wrong? Numerous examples exist for precision medicine in health and disease, but what is needed to help patients on Earth and future astronauts is more *predictive medicine*. Can a scientist actually engineer something and predict what happens? That is the best test of knowledge.

To this end, the first draft of the 500-year plan was posted on our lab's website in 2011, which included many of the ideas in this book. It was also the first year we submitted the genome and metagenome proposal to NASA, where we had almost none of the information described in this current chapter. Most of the ideas that seemed impossible in 2011 have already become reality, especially the ease with which we can now edit and modify genomes and epigenome (the regulatory landscape of the genome).

But beyond the rapid advancement of science, this plan represents hope and belief in the long-term survival of humans. One of my favorite things about humanity is that we are the only species we know of that can actually create 5-, 500-, or 5,000-year plans, or comprehend any multigenerational plan. Almost all the people who will benefit from such a plan will be born after the death of the plan's creators, yet such plans get made and can serve humanity like an intergenerational Olympic torch, bringing the bright light of past and planned progress to keep hope ignited and eyes looking forward.

The rest of this book will lay out this plan, which addresses the technical, philosophical, and ethical framework for engineering genomes, ecosystems, and planets. While seemingly abstract and almost unbelievable in scope, this large-scale engineering effort is not our first attempt. Mars will, in fact, be the second planet on which we have performed planetary-scale measurements, modeling, and engineering. In 2021, we are doing this planetary-scale engineering on Earth to continue our survival and leave a better planet for the next generations, but, sadly, with scant coordination or planning. We need to do such planetary and biological engineering with far greater precision in the future to fulfill our species' unique role of Shepherds and Guardians. It is no longer a question of "if" we can engineer life—only "how." Engineering life now exists within our generation and will continue to be improved and utilized for generations to come, be it those who exist in 500 years, 5,000 years, or much further into the future.

Engineering is humanity's innate duty, needed to ensure the survival of life.

2

THE DUTY TO ENGINEER

Plans are useless, but planning is essential.
—President Dwight D. Eisenhower

ENTROPY GOGGLES

The need to begin planning our future is best exemplified with a quick thought experiment called "entropy goggles." The first thing you must do is look around. If you cannot see, then *think* around. Be aware. Imagine what the world will look like 100 years from now. Consider whether any stairwell, wall, ceiling, or physical entity will survive. Think about how long your favorite shirt or your favorite pair of socks might last. Consider whether any other nearby living creatures, like a squirrel, will also survive, or their progeny. Examine the technology that has brought you these words—on printed paper, on a glowing screen, or by audio; did it even exist 100 years ago? How might it change in another 100 years? Examine everything around you, on you, and near you, and think how it may change; be aware of these changes. At the end of this paragraph, stop reading or listening for a moment and examine your surroundings. Imagine a world exactly 100 years ahead of you. Do this now.

Upon returning to these words, you might find the present time much more relevant and comfortable, but this is the first sensation that must be questioned. This comfort with only your own life, in your own time, is an unnecessary limit; it is only the beginning of the human mind's capacity. There is so much to imagine: change, decay, your own imminent demise, evolution, movement, and transience, but you can take solace in knowing that this time, here and now, exists. So, in terms of immediacy, this moment may be more comfortable than any point in the future. You may be startled by the knowledge that so much will change. But this is good.

As a conscious mind, you just utilized the ability to envision a world far ahead in time—100 years—one that you very likely will not experience. This ability to project a vision beyond the near future is unique to humans, as far as we know. This ability enables our creativity and ingenuity, as well as our unique awareness of our own frailty and that of all life. We share a lot with other animals on Earth—the drive to survive, the instinct to fight-or-flight when faced with danger, and the urge to protect our offspring (with the exception of a few animals who eat their babies). But we are unique. We are the only species with an awareness of extinction. Seeing such a threat of extinction, mapping its rates and trajectories, and worrying profoundly about children we will never meet who exist in a faraway future, is the first and necessary step toward understanding extinction risk and avoidance. Extinction is only a risk—and *not* a certainty—if we take action.

Not to use our unique ability to care for the lives of future generations would inevitably result in the loss of all other species' unique attributes. Each species has its own abilities, but ours is the only species that can safeguard the incredible diversity of Earth, all of the life in the universe. This ability can go even further, even to the preservation of life we will undoubtedly find outside of Earth as we explore into the continually larger universe, with countless other suns. If you only possess comfort with your own life, and not comfort with a vision ahead, then you unnecessarily limit yourself. Our life spans should *never* be a limit on how far we can think. Our comfort zone can, and should, exist long past the next 100 years. This ability to think far, far ahead as individuals, and as a species, is an extraordinary gift—a gift that must be preserved, protected, and used before it disappears forever.

THE BLUEPRINT

Importantly, the ability to plan this far into our future was not feasible before now. The current era of humanity is different, both in *degree* and *type*. First, the degree: Before the late 1900s, we lacked the fundamental tools to even begin to describe a plan which enables the preservation of humanity through expanding to other planets. We barely had a sense of the large diversity of the biological systems on Earth—which will undoubtedly contribute to our ability to live in new environments to which we have never been exposed. Before the past few decades, we did not even know what comprised genes, let alone how many existed in human cells or those in other species. Now, we know the dynamics of thousands of genes, and we have mapped them and their functions (functional genomics) across many species. We have more genetic data than ever, and these data are rapidly growing, along with the continual production of more data in related scientific, cultural, technological, and computational fields.

The second shift in this era is in its *type*: exploring other worlds is now possible. However, the concept of sending humans to the moon or Mars was pure fantasy until fairly recently, such as in 1906, when the first plane become airborne. The first humans landed on the moon in 1969. The first spacecraft to leave our solar system was only in 2004 (Voyager 1). Whereas airplane flights were once rare, now we have tens of thousands of flights around the world departing by the minute. Similarly, before we only had a few spacecraft in flight, but now we have an ever-expanding list of countries planning lunar, Martian, and even longer missions (figure 2.1). Current plans even call for a small helicopter (Dragonfly) to be present on the surface of Titan in 2036 and boots on Mars around the same time. The twenty-first century represents a unique timeframe for humanity, both in its *type* of progress as well as its *degree* of planning.

These missions exemplify how humans have the ability to think 100, 1,000, 100 million, or many more years further ahead. If we want

Planned Missions by Celestial Location

Exoplanets
- Starshot [Launch] ——————————— [Arrival] [Signal Returns]
- [Launch] [Complete] ARIEL (Atmospheric Remote-sensing Infrared Exoplanet Large-survey)

Outer Planets
- [Launch] ——————————— [Complete] Ice Giants
- [Launch] [Arrival] Dragonfly
- [Launch] [Arrival] Europa Clipper
- [Launch] [Arrival] [Orbit Ganymede] JUICE (Jupiter Icy Moons Explorer)

Asteroids
- [Launch] [Complete] Fast
- [Launch] [Arrival] [Completion] Psyche
- [Launch] [Inner–Main Belt] [L4 Trojan Cloud] [L5 Trojan Cloud] Lucy
- [Didymos Impact] [Hera Arrives] DART (Double Asteroid Redirection Test)
  [Launch] [Hera Launch]
- [Launch] [Return] OSIRIS-Rex

Mars
- [Launch] [Return] Mars Base Camp
- [Launch] [Orbit] [Return] Martian Moons Exploration Probe
- [Launch] Mangalyaan 2
- [Launch Robotic Ships] [Launch First Crew] Starship
- [Launch] [Arrival] Rosalind Franklin
- [Launch] [Arrival] [Sample Retrieval] Moon Global Remote Sensing Orbiter & Small Rover
- [Rover Arrives]
  [Rover Launch] [Retrieval Launch] [Retrieval Rover Arrives] [Return to Earth] Mars 2020 (Perseverance Rover)
- [Launch] [Arrival] [Completion] Hope

Moon
- [Launch Test] [Launch Crew] Russian Lunar Mission
- [SLIM] SLIM (Smart Lander for Investigating Moon)
- [Chang'e 5] [Chang'e 6 & 7] [Chang'e 8] Chang'e
- [Artemis 1] [Artemis 2] [Artemis 4] [Artemis 6] Artemis
  [Gateway] [Artemis 5] [Artemis 7]

Low-Earth Orbit
- [Launch] Gaganyaan
- [First Tour] Space Tours

2020   2030   2040   2050   2060

◇ NASA   ◇ Non-US Government   ◇ Non-Government

2.1 Near-term missions for space exploration and human settlement: Missions are grouped by locations, including low-earth orbit (LEO), the Moon, Mars, asteroids, outer planets within our solar system, and exoplanets. Years are given on the bottom and mission details are highlighted on each line. (See color plate 8.)

the diversity of humanity (be it in the form of music, art, science, literature, engineering, dance, and or anything else) as well as the diversity of life that has ever been, or is currently, on Earth to persist, we need to expand the catalog of Earth's life beyond the only home it has ever known. We should not abandon our current ship—the Earth—but should instead increase the number of ships on which we live. Currently, all known life exists on a very fragile raft adrift in a vast ocean of the universe's threats of extinction. Our responsibility to preserve life extends beyond our own, to a broader comprehension of the organisms that we ingest or utilize and of how other life-forms interact and sustain each other; this extension is often called the "metaspecies," "pangenome," or sometimes the "holobiont."

A vision of humanity as Shepherds should leverage our *unique* abilities as a species to preserve life and place us as careful Guardians of all life. For life to do anything, it must first exist. Any human dream, construct, ethic, art, manufacture, invention, creation, poem, synthesized molecule, or thread of fabric can only be made if we are still alive to make it. Regardless of your priorities and goals, you must *exist* to bring them to fruition. Even if you have no goals, existence is a prerequisite to holding an empty cup of ideas. Thus, we have a responsibility before us; we hold a heavy, hard weight in our collective palms. Only our own eyes can see the danger on the horizon, and only our actions can save the life we see around us.

Assuming this responsibility for current, past, and pending life-forms, we will then need a plan that will expand beyond the life span of our first sun, which gives us a maximum of 4–5 billion years. But even that maximum is highly unlikely to be the longest timeframe, given historical precedent. Any number of world-ending, catastrophic events can happen before then. For example, another asteroid could smash into the Earth and—as happened to the dinosaurs—we would all be obliterated. Based on our current planetary science estimates, we are actually past due for a planetary-scale catastrophic event—without even taking into account all of the harm that humans have inflicted upon our home planet. However, at the maximum, we have about four billion years; a finite time.

Our current sun will eventually run out of fuel and destroy all the inner planets: Mercury, Venus, Earth, and Mars will all be charred to a cinder. This inevitable red giant phase of the sun will obliterate everything that has ever been created, learned, or understood on this planet, notwithstanding the radio waves and other electromagnetic radiation broadcast since the early 1930s. Unless we find a way to somehow stabilize our collapsing sun, all of our technological, artistic, scientific, and cultural creations—indeed *all* of Earth's creations—will be destroyed if they stay here. We need a global plan to expand beyond this planet.

The *implementation* of any global plan would not be possible without the *means* to support both global coordination and communication. The internet and advanced forms of transportation have brought us this world only very recently. The constant and ubiquitous intercommunication of internet-connected devices is something most of us now take for granted. This connectivity has also led to an expanded industrial revolution and new "information age" that can rapidly build or destroy entire cities and countries. Somewhat like a drunken toddler with a flamethrower in one hand and a nuclear detonator in the other, we have emerged with an accidental empowerment to influence our entire planet's atmosphere and country-scale ecologies. These powers were born in the womb of cheap high-carbon energy sources, which now threaten our entire planet. As worrisome as it has been to watch the CO_2 levels on our planet rise, this may have been an inevitable progression of civilization. Now, with the ability to understand how these technologies can negatively impact our planet, as well as technologies to monitor and disseminate them (with global communication), we not only can fix the damage we have inflicted on Earth, but can begin to build better, cleaner civilizations on this world (and others) and ensure our own long-term survival, as well as that of any other species.

This key principle—the survival of as many of the life-forms and molecules as possible—is a new kind of ethics, a molecular and genetic ethics, which gives a purpose and duty to this idea of preservation. This is the highest (deontogenic) duty because all else depends on it.

DEONTOGENIC ETHICS

Deontogenics is a new kind of ethics originating from deontology (from the Greek *deon*, meaning "obligation or duty," and *ology*, meaning "study") and *genetikos* (from the Greek meaning "genitive or generative"). Deontogenic ethics is based on two simple assumptions. First, assume that only some species or entities have an awareness of extinction. Second, assume that existence is essential for any other goal/idea to be accomplished—in short, *existence precedes essence*. Therefore, to accomplish any goal or idea, sentient species (currently humans) need to ensure their own existence and that of all other species that enable their survival. Any act that consciously preserves the existence of life's molecules (currently nucleic acid-based) across time is ethical. Anything that does not is unethical.

Deontogenic ethics is related to, yet distinct from, deontological ethics, such as that formulated by Immanuel Kant. He argued that the morality of an action is based on whether that action itself is right or wrong, regardless of the outcome. Kant's "categorical imperative" asked people to think, before taking any action, "What if everyone did this? What if my action were suddenly a maxim for everyone? What would the world look like?" Deontological ethics is often seen as being in conflict with utilitarian ethics, such as that of Jeremy Bentham and John Stuart Mill, who aimed for "the greatest good for the greatest number." In utilitarian ethical frameworks, the outcome and consequences are usually more important than the action itself.

But utilitarian ethics also faces challenges of quantification and application. What is good, and how is it measured? What if there are situations that are technically "better" but actually worse for the average person? Derek Parfit wrote in his book *Reasons and Persons* about a "repugnant conclusion" of applying some of these utilitarian frameworks. For example, it would technically be "better" to have a large population with lower average happiness versus a smaller population with a higher average happiness. Another established ethical principle asks what would be "fair," regardless of which body you might be born into (e.g., rich or poor, powerful or meek), and was proposed as the "veil of ignorance" by John Rawls. Yet preceding *all* these discussions, debates, and frameworks is the need to exist in the first place. Thus, preserving the existence of life is the highest duty, a deontogenic ethic, since it necessarily precedes all others.

Deontogenic ethics includes four simple pieces: (1) consciousness must exist to be used; (2) long-term survival depends on plans to extend beyond the solar system in which our species originated; (3) long-term survival depends on the metaspecies, but is not only for the metaspecies; and (4) the needs of the metaspecies and conservation of their responsibilities may supersede individuals' needs or wants. While this may seem an affront to liberty—to remove choice from a conscious entity—it is worth noting that we already remove many aspects of choice with other "molecular infringers," including folic acid fortification in wheat and flour, iodine in salt, chlorine in water, and mandatory vaccinations. In all cases, the decision is made about the allocation of a resource before a person reaches that resource because it is for the greater good and survival of all other humans. It is the ethical action.

This deontogenic ethical framework can fundamentally change the view of many items, organisms, people, and cultures on Earth. For example, there are still some Indigenous peoples in Brazilian, Ecuadorian, and other rainforests who have never been contacted by the modern world. They represent a primal, sacred state of humanity that is worthy of study, preservation, understanding, and also an opportunity to find new practices, language, culture, and molecules. But leaving autochthonous cultures in the remote, dense jungles of the Amazon is actually a death sentence. Even if a tribe has achieved a perfect, serene, and warless human society, they are doomed. If they are still here in a few billion years, when the planet is engulfed by the sun, then all their knowledge, culture, language, and history will be permanently erased. Thus, the desire of some cultures to remain isolated from the rest of humanity, while perhaps reasonable in the short term, is wrong in the long term, and essentially results in premeditated group suicide.

This critique of a "suicide society" applies not only to isolated populations, but to the rest of the world as well. Humanity's looming suicide, rather than that of just one society, is on a planetary scale and applies to all societies and cultures if we do not get off Earth at some point in the future. It is worth noting that the combined knowledge

of our consciousness, geology, and astronomy makes this indeed a suicide, rather than what would have otherwise been called an extinction
or "accident," because it is fundamentally *known* and *preventable*. It is
equivalent to sitting on train tracks, knowing that a large train is coming down the tracks to destroy you, and just waiting for it. The technological, intellectual, and engineering challenges for leaving Earth and
settling on other planets are large, but not insurmountable, and can be
overcome if the volition and resources are dedicated to the effort across
generations. Not to act on our duty to the metaspecies is a failure of
deontogenic ethics. It is a failure of the duty to our own species and all
others. So why would anyone say no?

CRITIQUES OF LEAVING EARTH

Resistance to the idea of expanding to other planets often comes along
a few, predictable paths: solipsism, prioritization, indifference, and
futility. On the first path, of selfishness, someone may say, "I will be
dead, so why should I care?" This position is shortsighted, and it also
violates Mill's ethics, Kantian ethics, Rawlsian ethics, and deontogenic
ethics; plus it assumes no responsibility to others who live afterward.
It is not a sustainable position. If everyone held this view, the person
speaking such a sentence would have a limited or nonexistent world in
which to live and thrive, or may never have been born at all.

The second path of resistance involves the prioritization of goals
and is usually phrased as: "We have poverty, disease, and other press-
ing issues that we need to attend to first." However, conquering disease and other societal problems can coincide with working toward the
overarching research and development goals of settling new homes in
space. To force a choice of one or the other is a false dichotomy. The
US Space Program in the 1960s managed to get boots on the moon,
as well as grow the economy substantially without having to choose
doing only one, so striving to achieve multiple aims at once is clearly
possible. But at the time, NASA devoted 4.4 percent of the US gross
national product to this goal, while today the percentage is tenfold
less—a scant 0.47 percent. Nonetheless, we can chew gum and walk at
the same time.

The third path is just avoidance. It might sound like, "I'm not directly related to this work, so I don't need to do anything." Here, too, this view is misguided. In all countries with space programs, taxes support the work on space biology, rocket engineering, flight logistics, and astronaut facilities. As such, the citizens of all these countries are already involved. Moreover, citizens don't have to be intimately involved with a national or international project to support or appreciate the benefits of such a project, such as peacekeeping forces from the United Nations or work on global disease tracking from the World Health Organization.

The fourth path of resistance is based upon the longest-possible view of the universe and so can lead to indifference. One might say, "If we leave this sun, we will just have to leave the next sun, and then again, again, and again. Where does it end? Won't we all just die anyway when the universe ends?" This contention is based on the second law of thermodynamics, which includes the biggest system of all—the universe. The fate of the universe depends on its total energy and matter, yet "regular matter" makes up only ~5 percent. The density and activity of dark matter (27 percent) and dark energy (68 percent) are the biggest factors. Nonetheless, after trillions of years, the plan of simply expanding to more and more solar systems will not be our best choice. After millions, billions, or potentially trillions of years of interstellar travel, it will eventually be as trivial as it is to go from NYC to Paris today. After visiting many stars and gaining first-hand experience and data on how stars form and die, we may be able to engineer our way through this problem, as we have countless others. But can we make it to the end? What is the end?

The current understanding is that the universe will end in one of two ways. First (most likely), it could be from the universe expanding endlessly, called the "big freeze" of the universe. This is when planets continue to drift apart, then cells, then molecules, then atoms, and then, eventually, even the very, very old (10^{35} years) protons themselves will be too far away from each other to interact. The other potential avenue of our universe's demise could be in the "big crunch," where eventually the universe will stop expanding and then begin to fall back in on itself. In this scenario, dark and visible matter/energy of the universe becomes dense enough to drive all mass to continuously move closer together and possibly lead to a new big bang (more in the last chapter).

Outside of the technological challenge of solving these two scenarios, which are currently unsolved, there is also an ethical question. Should we restructure fundamental atomic and physical properties of the universe in order to preserve life? What if the universe has already had a big crunch, or several? Perhaps such cosmic cycling is what preceded our own "big bang," and life could arise again in the new universe. Moreover, life could arise in a better form in the new universe, and if we stopped this from happening, we could be doing harm to life's prospects. How can we know the impact of our decisions here in the long term?

Here, the question is easily resolved by deontogenic ethics. We know that humans are still the only species with the knowledge of extinction, and thus, we have a duty of stewardship of our own species and that of others. A worse outcome than a universe with imperfect life is a universe with no life, since then there is nothing to protect or maintain the universe, as well as its life, and that is a risk that is too great to accept. The scale of the hubris does not obviate its necessity. If the intent is to preserve life, which has the ability to preserve the rest of the universe, and if the act of doing so does not itself harm what it is trying to protect, then ultimately the most moral thing to do is to act.

Moreover, life likely would not need to end. After *that long* of a time period, humans (or our derivative species or robot brethren) will be substantially different and probably much more technologically advanced. It may even be possible that by that time (billions or trillions of years from now), we will have the ability to measure, manipulate, and use dark matter, spacetime, or other tools to enable changes to the structure of the universe itself. This new era that is trillions of years ahead may include spacetime folding for long-distance travel and matter manipulation at the scale of entire stars, galaxies, or even scales across the universe.

We may have to confront a choice of either letting the universe die, hoping that life will be born again, or actively preventing its death and restructuring it to preserve life. Considering molecular and deon-togenic ethics, we only have one choice. If we are to preserve all life, we will have to restructure the universe itself in order to survive it; our duty is to engineer.

This duty to engineer and protect life is the only duty that is imme-diately activated upon comprehension. Other duties, such as those to one's children, country, family, or religion, can sometimes be delayed, switched around, or even abdicated. But the duty to the universe, and to all life, is something that will always hold true for us, as well as for any subsequent species or entity (made of any matter) that realizes this responsibility. This is a duty that enables, for the first time, a piece of the universe to keep the universe's creations sustained, protected, and thriving.

3

PHASE 1: THE LANDSCAPE OF FUNCTIONAL GENOMICS (2010–2020)

> What is the philosophy of the gene? Is it a valid philosophy? There has been too much acceptance of one philosophy without questioning the origin of this philosophy. When one starts to question the reasoning behind the origin of the present notion of the gene (held by most geneticists) the opportunity for questioning its validity becomes apparent.
>
> —Barbara McClintock

We have two options to enable the transition of Earth's life to start living away from its protection (stable gravity, magnetosphere, atmosphere, pressure), much like a bird leaving the nest for the first time. One possibility is we simply allow evolution to gradually select for characteristics required to survive on these new planets—natural selection. This is basically a "sink or swim" approach to life's survival, except with no lifeguards and with bricks tied to your feet. Though this type of selection could gradually work, it would take an extremely long time, or may simply work too slowly to even be viable. On the harshest new planets, any Earth-based organisms might die before they're even capable of reproducing.

Our second option to enable Earth's life to live on other planets is to preemptively direct this genetic process, so that the life we send is already capable of surviving in its new home. More complex, yes—but also more humane. However, to do this we need a better understanding of the "functional" elements of all genomes in the metaspecies so that we can better engineer and protect them. To make this a reality, we must first comprehensively map the functional elements of the human genome, our metagenome, and other intermingled species (the holobiont or metaspecies), so that we have an exhaustive list of what must be protected, what can be edited, and what abilities are even possible to engineer based on Earth's current biological catalog. Before we build genomes from scratch (as we will discuss later), we need to know what we even *can* build.

Though our own genome is the most well-studied genome to date, we have yet to completely unravel all of its mysteries. How many genes do we have? What are they doing? How are they regulated? Are there some "do not disturb" areas of the genome alongside those that can be more readily modified, altered, or manipulated? What about for other species?

Answers to all these questions are essential for the subsequent steps of biological engineering. Beyond this, we need a better understanding of how life responds to space travel through adapting the methods employed in the NASA Twins Study (chapter 1) to more astronauts and species. As of the time of this publication, there have only been 570 human beings who have ever been in space, 100 km above the surface where the atmosphere thins away and the sky becomes black. Once space tourism launches (e.g., SpaceX and Blue Origin) and newer space agencies begin regular human flights (e.g., Israeli and Indian), the amount of data generated will increase exponentially. At the same time, the work on animal and plant models (at NASA's GeneLab, as the largest and best example) will also increase. A cotton plant has even already been grown on the moon thanks to China's Chang'e-4 mission. Lunar socks made from this rugged cotton are not yet available for order, but it is only a matter of time.

But while all this momentum is exciting, the question still remains: What should we engineer? First, we must define substrates and some basic knowledge to understand what we *can* engineer. To capture this perspective, we must first build a context, from the genome, to genes, to cells and tissues, to the whole body, and then to the entire ecosystem. Only once we have all the molecular mortar and genetic bricks, can we build the biggest structures, including ones that reach the stars.

GENOMES

The fundamental units of a genome (also called the genetic code or DNA) are genes. These genes are built from the four nucleotides (A, C, G, T), which can be "transcribed" (think reading a text out loud) as RNA (A, C, G, U)—commonly written simply using letters of the alphabet. DNA is transmitted through each generation and is the primary hereditary molecule. When it becomes active, it is transcribed into RNA as a "messenger" molecule to carry out the function of the DNA. RNA may act as blueprints to be "translated" (think spoken words to action) to create proteins (coding RNA) while other RNAs are already active molecules by themselves (noncoding RNA, or ncRNA), which can regulate other processes inside a cell.

These proteins and ncRNAs create the framework for how cells grow, adapt, and signal to other parts of the body. They compose the active components of the genetic code that grow fingernails, replace lost blood cells, digest food, release biochemical euphoria during orgasm, guide sleep, and enable the human body not only to survive but to thrive. The collection of all the genetic code for a single organism is called the genome; if it spans a range of organisms considered together, this is called the metagenome.

While the human genome may seem large at ~3.1 gigabases (3.1 billion bases), it is actually not the largest one. Most plants have far larger genomes (10–30 gigabases), and the largest known genome is in the euglena, which has 120 gigabases. Perhaps the most striking aspect of the human genome is not its size but rather the amount of coding RNA relative to noncoding areas of the genome (noncoding RNAs and other areas between genes). Many ncRNAs define very specific types of cells, as shown by work from Ari Melnick, Manolis Kellis, Mark Gerstein, John Mattick, Christina Leslie, and the ENCODE Consortium. Indeed, this is the one aspect of the human genome that may represent its most unique features relative to other species, with only ~2 percent of the genome encoding for proteins. In contrast, bacteria have 99 percent of their genome for protein products, yeast has 80 percent, and most other organisms have around 20–30 percent. But . . . why? How many genes are there, and can we predict what they are all doing?

GENES

The DNA and RNA present in each cell do not simply use a four-letter code to provide the vast array of instructions and innate adaptive responses to the world around them. Rather, there are discrete, functional elements of DNA that serve as the levers of function, called genes. In the 1940s and up until the 1970s, the predominant theory was that one gene would perform one specific function ("one gene, one enzyme") because this is what was observed in bacterial systems, and this even led to a Nobel Prize for George Beadle. This prior work has led to a popular conception of genes such as the "intelligence gene," the "cancer gene," or the "height gene."

However, in human and other multicellular organisms, gene regulation is not so simple; rarely does one gene perform a single, simple function or drive a trait (a.k.a. phenotype) or have a sole effect on the risk of disease. Almost all genes are active in more than one cell type, across different tissues, and/or at different times of development, and this "one gene—many functions" principle is called pleiotropy. Because many genes have a pleiotropic function, it can be very difficult to point to just one gene that is driving one specific phenotype. For example, a protein that can serve as an enzyme for one chemical reaction can serve as scaffolding for another purpose inside a cell; the same gene that controls insulin metabolism (PI3K) can also be critical in a cell's response to chemotherapy and whether or not cancer becomes metastatic.

While the human genome's size has remained about the same ever since the divergence from our last common ancestors, chimpanzees and bonobos, around 6 million years ago, our knowledge of genes and their regulation has only been clarified in the last ten to twenty years. In the late 1990s, the number of genes in the human genome was a subject of great debate, with some positing that there would be as many as 120,000 genes or as few as 20,000 genes. The distinction between coding RNAs and noncoding RNAs was not yet well understood, and many people had an anthropocentric view that humans appear complex, and as such, we should have far more genes than a fruit fly (13,000) or a worm (20,000) has. Yet when the first full draft of the human genome was released in 2001, we ended up with only 25,000 genes.

Excitingly, the pace of discovery of new genes within the human genome has not slowed since 2001. Work from many researchers in individual laboratories, as well as across large consortia such as the Encyclopedia of DNA Elements (ENCODE), has been quickly uncovering new genes within the 3.1B letter book of life, averaging about 1,000 new genes every year. The number of genes increased to beyond 60,000 in 2020, and more are likely to be uncovered in years to come. Although the number of protein-coding genes, which made the enzymes, protein complexes, and amino-acid-based functional elements in our cells, has stayed relatively constant at 20,000 genes, the number of noncoding genes that are clearly defined is continuously increasing (figure 3.1). This means that some of the most critical genes for adaptation to space-flight may still be awaiting discovery.

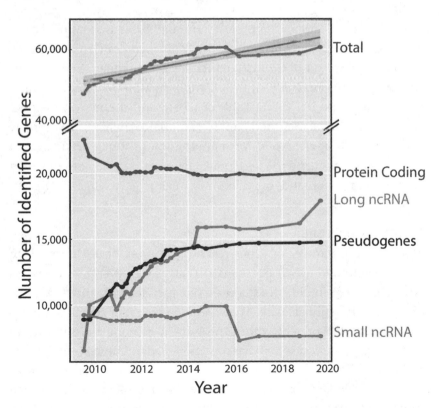

3.1 The number of identified human genes over time: The GENCODE gene-annotation count for each category of genes, including the total, protein-coding, long noncoding, small noncoding, and pseudogenes.

These discovery efforts are still ongoing because the substrate of DNA just holds the information of the human genome, whereas the active form (RNA) can be exquisitely specific to each cell, tissue, or time of development, and, thus, can take extraordinary effort to find. For example, fetal hemoglobin is the molecule that fetuses use to process the majority of the oxygen in their bodies, but this gene is turned off around birth and usually never seen again. To address this challenge of spatiotemporal complexity and genes that may be very rarely expressed (in time and/or space), several ongoing projects are building an atlas for each cell type in the human body, including the BrainSpan project led by Nenad Sestan and the Human Cell Atlas (www.humancellatlas .org) led by Aviv Regev, both of which can help discern how may genes are truly present in the cell from that very first cell.

But discovering genes is not just an endeavor for the human genome. There are several large-scale projects and databases that store genetic information as it is growing over time. This includes GenBank (more on this when we get to CRISPR), which keeps all sequence data generated around the world, as well as the European Molecular Biology Laboratory, KBase, the DNA Data Bank of Japan, and the newest member of the genetic database, the Chinese National GeneBank. Two of the largest projects in the world for mapping the genomes are the Earth BioGenome Project and the Vertebrate Genomes Project, both of which are discovering thousands of genes every week across a range of habitats.

GENETIC CHANGES

Whenever the number of genes reaches the final, "true" number, or a near asymptote, it will not remain that way forever. Life is always evolving, and even old "dead" genes, like pseudogenes, can "return to life" and be functional again. These genes are relics of genetic information that are still present within the human genome and serve as an evolutionary palimpsest for what has happened to human biology over millions of years. Our genome is essentially an old piece of paper with billions of overlapping scribbles and edits of evolutionarily selected "notes," which we now have the ability not only to read, but to see *what* changed and *how*. One process is called exonization, where a portion of a gene that is not currently transcribed into a protein becomes mutated and then turns into an exon, which can be used as part of a new RNA or protein element. Also, almost all genes undergo some splicing, with internal elements of the genes being mixed and matched to create a new function. This splicing process can occur in the case of disease (such as myelodysplastic syndrome), sex determination for male versus female, or specific immune responses.

Beyond the genesis or recycling of a new genetic component in the toolkit of life, there is also a simpler way to get new genetic functions, which can just occur with "selection." Some instances of rapid evolutionary selection have occurred within only the past few hundred years. Most visible in daily life is the ability for humans to digest and process the lactose in milk as adults (lactase persistence), an ability that normally is absent in mammals beyond the infant years. Also, the genes that enable free diving at great depths and larger spleens have been selected for in the Polynesian islands; hence, these islanders can now dive deeper and longer than other humans. Finally, there is some evidence that genetic selection (for the EPAS1 gene) in Sherpas and Himalayan climbers has led to them being better adapted at life in high elevations. Evolution has already given humanity some recent adaptations, which only took a few dozen generations.

GENETIC REGULATION

While the genome (all of your DNA) and transcriptome (all of your RNA) define the cell's basic building blocks and potential, their regulation is controlled by additional molecules that sit on their bases and are collectively called the epigenome and the epitranscriptome, with "epi" meaning "above." There are hundreds of chemical marks that define when, how, and where DNA and RNA are deployed and used in cells. They can range from very small chemical changes—such as DNA methylation, wherein only four atoms (CH_3) are added to a cytosine (letter "C") in DNA to help control a gene's function—to large changes to the DNA or the proteins around which it is wrapped.

For RNA, the same principles apply, with slight chemical modifications like methylation modulating the function of a given RNA, which was first defined as the epitranscriptome by our laboratory and others in 2012. There are now over 115 known RNA modifications, which span all domains of life, and which represent a remarkable plasticity of RNA, just like DNA and the epigenome, in controlling the state, localization, translation speed, and stability of RNA. Just as DNA cannot be imagined without the context of its modifications and packaging, the same is true of RNA, too.

Almost all RNA-based viruses have modified RNAs, including the human immunodeficiency virus (HIV), Zika, and hepatitis C virus (HCV). Work from Dr. Stacy Horner's laboratory at Duke and our own has shown that these modifications change how fast the viruses grow, are released, and interact with host cells. RNA modifications have been seen in almost every tested organism, including viruses, plants, bacteria, fungi, and animals. It is now understood that, like the epigenome, the epitranscriptome serves as a set of hidden "levers" that control the function of RNA. These levers are potential substrates for future cellular engineering.

CELLS

Every multicellular organism begins as one cell, which contains all of the intricate instructions to synthesize, organize, and regulate not only this cell but the development and maintenance of all cells that will inevitably comprise the organism. All of these instructions are encoded in the first cell's DNA. This underscores the complexity of the genome and how each cell's expression must be controlled in specific ways depending on its function. The cells hailing from each tissue in the human body (e.g., muscle, lung, heart, liver) harbor a unique epigenetic signature, which enables the maintenance of tissue-specific functions through the control of gene regulation, as just discussed.

Our knowledge of the total number of unique cells, or cell types, is still growing. Previous estimates put the number of unique cell types in the human body at ~300, but new estimates from the Human Cell Atlas have shown that we may have thousands of cell types and subtypes, each harboring a unique function for a specific physiological state or response to stimuli. But even cells of the same cell type will not be identical. A cell's "presentation" of molecules on their surface can radically change depending on internal variables such as genetic mutations or altered states of their epigenome, transcriptome, and proteome, as well as external stimuli including drugs and interactions with other cells. This novel presentation is most pronounced with a neoantigen, when a cancer cell creates an entirely new molecule on the surface of a cell. Given its unique presentation, which wouldn't be found in normal cells, this offers a unique target for safer cancer therapies.

The human body has about 30 trillion human cells plus another 30–40 trillion bacterial cells, for a total of about 70 trillion cells. If your body were a democracy, the human cells would often be the *minority or equal* party. You (as a human) would never win an election. Your loss of control would likely result in you rolling around in the soil or lying in a bathtub full of yogurt, which I do sometimes on Sundays. Regardless of how you spend your Sundays, there are a lot of microbes in, on, and around your body. There are in fact so many microbes that they compose the bulk of the cells on Earth. This is a humbling and exciting statistic, and one which is vividly apparent for anyone who has ever had explosive diarrhea.

While bacterial genomes are smaller in size (2–10 megabases vs. 3.1 gigabases for human), their biochemical activity is as important as, and sometimes more important than, the human component. Estimates by Lee Hood showed that 36 percent of the small molecules in the human body are either made by, or processed by, the microbiome. And about 25 percent of drugs that are designed for human disease can also affect the growth and biology of the body's microbial cells. As such, a treatment for a disease is never a treatment for one person; rather, it is a treatment for all cells across all kingdoms/domains of life.

Yet this is only one facet within the large complexity in studying disease and predicting treatment response. Our continual understanding of the true complexity of biology has enabled predictive modeling and patient-specific customized therapies as the new medical paradigm. Centers for "precision medicine" and "personalized medicine" have become common at hospitals and medical centers around the world, with the goal to deliver the right treatment, to the right patient, at the right dose, and at the right time. Precision medicine has led to extraordinary breakthroughs in customized treatments for cancer, especially leukemia and lung cancers, where the "Achilles' heel" of the cancer can be found and exploited, such that it will kill only the cancerous cells and leave the nonmalignant cells alone. Also, work in infectious disease and "metagenomics," or all DNA regardless of species, has enabled "precision metagenomics," which can enable patient-specific antibiotic matching, or the discovery of mysterious species involved in the patient's disease, from which we can continually learn. Work in metagenomics has also led to unintended discoveries of microbes *within* tumors that possess an ability to process and metabolize chemotherapeutic agents and lead to resistance to therapy. Thus, some tumors may need to be treated first with antibiotics to kill the bacterial cells, and only then can the tumor cells be targeted. These treatment regimens represent a complete shift in medical treatment from preceding years and a new view of "cross-kingdom biology" for medicine. To be a good human geneticist today, you must look beyond human DNA, since cells from all kingdoms of life and DNA from all nonhuman cells can influence every human.

Indeed, if you are a human geneticist and have only examined human DNA, you are a bad human geneticist. It is simply no longer sufficient to only be concerned with one field, be it microbiology or oncology, and have that suffice as if these biological processes act alone. Although most biology books show the three kingdoms of life as separate branches in an evolutionary and historical perspective, this view is misleading in both a clinical and biological context. Humans are multikingdom entities (bacterial, viral, fungal, other cells, and DNA), with a continual stream of information flowing between each domain of life. As such, we must be viewed, measured, and modeled as a dynamic, cross-kingdom network of life. Proper modeling of disease, health, and biology should, therefore, be agnostic to the cell, species, or kingdom of origin but instead focus on all molecules made by any cell within a system, which can, and inevitably will, interact with other molecules. This perspective is also essential to how genomes should be designed and how they will be sent to other planets.

METAGENOME MAPPING

So, what is the extent of the required information for a model that is inclusive of all potential molecules from any species? The total number of species on the planet is estimated to be in the trillions, and yet our mapping of them, and their interactions, has only just begun. There are large-scale efforts to map microbial and environmental DNA in many places around the world. These efforts include the Earth Microbiome Project, which is discovering new organisms in a myriad of habitats to map their diversity and functions, as well as the Extreme Microbiome Project, which looks for extremophilic organisms and examines their DNA for metrics of adaptation (for more, see chapter 5). Also, the Metagenomics and Metadesign of the Subways and Urban Biomes (MetaSUB) project has sequenced thousands of samples from >100 cities and released its first data in 2020, from the work of David Danko, Daniela Bezdan, and the >150 city consortium. Although at time of release, these data only included cities found on Earth, there were still 4.3 million new genes found and >10,000 new viral species. These findings underscore the amount of pure biological discovery that is left to be uncovered with genomic sciences in Phases 2 and 3 of the 500-year plan (including creatures on subway poles).

This entire system of bacterial, viral, fungal, archaeal, human, and other cells and molecules and their physiological and biological integrations spans the vast metagenome (all genomes). Engineering the microbiome and metagenome is an ongoing priority for NASA and other space agencies, also called the Microbiology of the Built Environment (MoBE), which was a field pioneered by Paula Olsiewski at the Alfred P. Sloan Foundation. All MoBE sites, from homes to subways to space stations, represent a currently "accidentally" engineered ecosystem for their inhabitants. But mapping and characterizing this has been historically difficult.

In 2018, the Earth BioGenome Project was launched, aiming to sequence and annotate ~1.5 million genomes of all complex life on Earth. As part of the European effort, the Darwin Tree of Life Project was also launched in the United Kingdom. These groups plan to work together to sequence and annotate all 66,000 eukaryotic species from the British Isles. Together with partners from a breadth of UK-wide institutions, museums, and universities, the Wellcome Sanger Institute serves as a hub for the sequencing of these genomes, and our operational pipelines will play a central role in delivering the generation of these genetic codes. This will create an extraordinary, first-ever, planetary-scale map of genomes. But that only encompasses everything below our feet—what about above us, in space?

Much of what we know about the space station's metagenome comes from Dr. Kasthuri Venkateswaran at NASA's Jet Propulsion Laboratory (JPL). Evidence of organisms rapidly adapting to space, including antibiotic resistance, has been observed since the early 2000s and has direct implications for crew health and safety. While other sampling efforts have indicated that the organisms are adapting, they may not necessarily become dangerous. As is the case with most things in biology, the impact is case and time dependent. Until 2015, it was not possible to monitor these changes in real time on the ISS, but then an idea for an experiment was born.

DNA IN SPACE

Ideally, one could just sequence the DNA or RNA of any organism of interest right on the ISS, but the machines to perform such tasks were too large. Most DNA sequencers, while extremely fast and high throughput (e.g., Illumina), were all too heavy to justify the cost of putting them in space. But, in 2012, Oxford Nanopore Technologies announced a miniature sequencer called the MinION that would fit in the palm of your hand and only weighed 0.3 kg. Then, in 2014, they gave my lab at Weill Cornell and others early access. In 2015, we made sequencing in space a reality. It turns out, all you need are reagents, a computer (even a tablet), the MinION, and some guts.

While planning the logistics for the Twins Study, I asked NASA if it would be possible to get a nanopore sequencer on the ISS, and they suggested I meet Drs. Aaron Burton and Sarah Castro-Wallace. It turns out they had already begun planning a mission along these lines, so we joined forces for the Biomolecule Sequencer (BSeq) mission along with Kate Rubins and Charles Chiu.

With this unstoppable team, the planning began. Our first test was to address a simple but important question: will the sequencer even work in zero gravity? What if we take it all the way up to space, but it doesn't actually work there? What if, for some reason, the sequencer works differently in reduced gravity than it does in regular gravity? Fortunately, there is the parabolic flight simulator, known as the Vomit Comet, which NASA uses for testing zero-gravity protocols, and you can imagine how it got its name.

On one of the Twins Study planning calls, Dr. Andy Feinberg mentioned he was going on the Vomit Comet to test some new positive-displacement pipettes to see if they work in zero gravity, and he asked if anyone had ideas for other experiments—my answer was clear (yes!). We had a perfect chance to test the sequencer in zero-G. We got all the supplies shipped right away and they boarded the plane, quickly jumping into the zero-G experiments. As Andy got started, chips and tubes were flying all over the plane, but he managed to maintain his composure and stayed focused on his task at hand, transferring samples and making sure the pipettes worked. Eventually, he was capable of loading the little sequencer, and, together, we showed for the first time that it was possible to sequence in zero gravity in 2015.

Next up: the ISS. Dr. Kate Rubins, a member of our BSeq team, a trained virologist who previously had her own research lab and had already done plenty of sequencing on her own, was the next astronaut selected to go into space—the timing could not have been more perfect. She safely made her way up to the space station in August 2016. NASA sent up a resupply rocket with all the supplies, and we coordinated to do the analysis between Houston, New York City, and the University of California, San Francisco, to make sure everything was working, in real time, on the ISS. The experiment worked flawlessly, and for the first time in human history, a person sequenced DNA outside of Earth, ushering in a new age of "space genomics" and what comedian Trevor Noah later called "space genes." We published the first genome and epigenome from these studies with members of the BSeq mission and other collaborators. This publication includes the first-ever genetic and epigenetic data from DNA that came not from Earth, but rather from space.

Within this new age of space genomics, future astronauts will be more self-reliant, a crucial requirement if you are stuck on another planet. If they are faced with unknown or challenging medical problems, such as an antibiotic-resistant microbe, they can sequence it and tell precisely what organism it is, and also find the best potential course of action. Further, as previously described, sequenced-based monitoring techniques will now be able to detect a variety of diseases early and monitor stress over time.

THE MOLECULAR BODY

All of the discussed biological layers—from the epigenome's regulation of what genes are transcribed into RNA, to the epitranscriptome's regulation of how these RNAs are processed and when they are translated into protein, to additional modification on proteins and the interplay of all of these molecules from cell types arising from vast areas of life including viruses, bacteria, and animal cells—together make up the "molecular body." When aggregated together, all of these elements fundamentally change how we view the human body. So, how can we easily monitor issues when they arise within such a complex system? Enter cell-free DNA (cfDNA).

A large number of small fragments of cfDNA are currently released from the cells which once held them (either intentionally or as cells die) and are taking a free tour around your body on the highways of your bloodstream. In healthy individuals, cfDNA is predominantly derived from the death of normal cells, especially blood-related cells, with additional contributions from other tissues. Notably, 1,000 to 10,000 genome equivalents of cfDNA can be isolated from just one milliliter of plasma, composed of human chromosomal and mitochondrial DNA, as well as fragments of viral, bacterial, and fungal genomes. Given that the cfDNA is essentially the "garbage scan" of the body, sifting through this garbage can identify what is happening in your body's molecular party and what your cells are throwing out—much like a garbage can full of beer cans or wine bottles after a long night of partying. The relative amount of cfDNA, and the sources from which these fragments arise, can, therefore, be monitored, offering an information-rich window into human physiology. In the 2010s, these tools rapidly expanded in applications ranging from prenatal testing and cancer diagnosis to the monitoring of infection, rejection, and immunosuppression after solid-organ transplantation. Cell-free sequencing of DNA and RNA (cell-free nucleic acids, or cfNA) in the bloodstream showed that it is possible to make a "whole-body molecular scan" based on those measures.

This work on cfNA has led to several new predictive measures that can be deployed from one blood draw. First, because cancer cells often have mutations that are distinct from the normal cells of the body, the higher abundance of these mutations in cfDNA can reveal the presence of cancer, perhaps even before the patient is sick. This concept has fundamentally changed how we can screen for early detection of cancer, including early pan-cancer detection cfDNA tests (such as ones from GRAIL) as well as ones from stool specifically to identify colon cancer (e.g., the Cologuard screening test from Exact Sciences). Second, a transplant patient receives a large number of cells from another person, and these different alleles (versions of genes) can be tracked in the recipient if an organ is being rejected and destroyed. In one 2014 study from Stephen Quake and Iwijn de Vlaminck, successful heart transplants could be correctly identified by the absence of the donor-derived DNA in the plasma, whereas those heart transplants that failed (organ rejection) showed an increase of the donor DNA as their bodies started to attack the foreign cells. cfDNA has further changed how we monitor sepsis patients, kidney transplants, and diseases based on the abundance of unexpected (microbial) DNA in circulation.

As discussed in chapter 1, it has been shown that mitochondrial DNA (mtDNA) spikes in the blood of people giving a speech in a crowded room. Other studies on mtDNA levels in blood during times of stress have shown similar results. A study from Lindqvist et al. on people who just attempted suicide showed elevated mtDNA in blood plasma, and this was consistent across many kinds of suicide attempts (drug overdose, hanging, wrist cutting, and some even in combination). Indeed, mtDNA peculiarities even showed up in the NASA Twins Study. During the first week in space, and again toward the end of the mission, Scott showed large increases in the amount of mtDNA in his blood, indicating significant immune stress from spaceflight including radiation, fluid shift, and environmental changes. This mtDNA spike represents a way to gauge the health of future astronauts too.

But with the metagenome and the holobiont in mind, monitoring can and has revealed changes across all species. For example, other work from the De Vlaminck lab at Cornell University showed that sequencing cfDNA from the urine of patients with urinary tract infections can reveal not only the type of infection, but the causative organisms as well. Further, treating cfDNA with a chemical called bisulfite can reveal its epigenetic origin, enabling us to see which organ in the body has been infected due to the differential cell type regulation of the epigenome, as previously described. This gives a simple, noninvasive method for profiling possible low-grade infections and also a means by which to examine the damage states of tissues and cells around the body. Every fragment of DNA has its own epigenetic origin story; we just have to listen.

SPACE DIAGNOSTICS

Precision medicine has become more precise over time, especially with the influx of data ranging from different disease states and molecular regulatory levels, leading to an improved ability to treat and monitor cancer, infectious diseases, and inherited diseases. As shown above with Dr. Rubins, these abilities expand beyond Earth, and it is now possible to perform continual genetic-based monitoring of astronauts in space. Similarly, machine learning algorithms have drastically improved on Earth in recent years and will undoubtedly play important roles in future space diagnostics. Some examples include the ability of image-segmentation algorithms to diagnose breast cancer, stratification of high-risk and low-risk patients from histology slides, identifying embryos with the best chance to implant (from Dr. Iman Hajirasouliha and Olivier Elemento), and predicting drug toxicities before wasting time and money running clinical trials. Beyond that, the use of "big data" in medicine has also pioneered new methods using electronic medical records and wearable devices to predict patient health trajectories. We have not yet generated a comparable "big data" repository of astronauts, but it will be necessary to predict risks and identify how to respond before they even leave Earth.

For example, work from fellow NASA Twins investigator Michael Snyder at Stanford University has shown that wearable devices for continuously monitoring heart rate and changes to DNA and RNA can reveal a risk of diabetes before it emerges, as well as Lyme disease before it becomes too serious. The same has been shown for COVID-19 from SARS-CoV-2. If you continuously monitor your physiology, it can serve as a self-sentinel. Overall, these methods and tools show the immense power for predictive medicine that may harbor a newfound capacity to stop diseases before they even start. If all goes well, they will soon be standard for astronauts on every NASA or other space-agency mission—enabling astronauts to use their own baseline data to identify changes within their "molecular body" and, if a therapeutic intervention is necessary, how and when to take it.

THE FINAL GENOME

One really surprising fact about the human genome is that we keep "finishing it," again and again, like a geneticist's favorite video game. Though it was first finished in 2000, with a big press release on the White House lawn with President Clinton, it was "finished" again in 2003, and then again and again. Every few years since 2003, there have been fixes, tweaks, and official new releases of the new human genome by the Genome Reference Consortium.

In fact, the genome was never completed in the first place. Even today, the 3.1 billion piece jigsaw puzzle that is the human genome has not been completely solved. Our current best mapping exists as chunks of hundreds of fragments of DNA, rather than a nice, neat set of tidy chromosomes. However, recent technological developments and computational methods have enabled a new era of finished genomes. This includes the complete end-to-end assembly of the human genome.

So, what's the problem? Why have there been so many revisions of the human genome? To return to the puzzle analogy: given current technology, the genome is chopped up into millions of tiny fragments and then read as strings of letters (called reads). Using computational methods, we then "align" these reads to each other by matching the same string of letters. Some regions of the human genome, and therefore the fragments from these regions, are highly unique, like corner pieces of a puzzle. But other areas are much harder to solve: effectively we have thousands of the exact same puzzle pieces, except we aren't allowed to use them as substitutes for each other. At the end of your chromosomes are highly repetitive "TTAGGG" sequences known as telomeres. Having your computer read this region of the genome out loud would effectively create a stuttering robot. Then, in the middle of the chromosomes, called the centromeres, there are additional repeating segments that are impossible to resolve if you only have short pieces of DNA sequences. But as the sequencers such as the Oxford Nanopore MinION and the Pacific Biosciences instruments have improved their ability to accurately read longer sequences, assembling the puzzle of life has become more tenable.

This is exemplified by work from Adam Phillippy, Karen Miga, Erich Jarvis, and others at the Telomere-to-Telomere consortium, who have shown that we can now construct a whole human genome from scratch. Even the X chromosome, the female sex chromosome, was only completed in its entirety for the first time in 2019. But now that one is done, the others will quickly follow. We now can map the location, phasing (position relative to other genes on the same molecule), and state of each of the genes, as well as what is expressed, what is modified, and what is the likely impact of functional elements, such as enhancers (distant control boxes for genes).

Thus, by 2021, the majority of the basic mapping of the human genome is complete. This has enabled the identification of almost all of the first-draft candidates of "do-not-disturb" genes, which we should not edit, remove, or modify. These areas are defined by large-scale efforts to understand human genetic variation, including the Genome Aggregation Database, the UK Biobank, and other population-scale sequencing efforts around the world, which also help with this map. These efforts provide the genetic data, medical data, and phenotypes for millions and millions of people, which can be used to identify mutations that likely have no negative impact on reproduction or quality of life. As an example, if you look at millions of genomes and 10 percent of them have a mutation, then that mutation is probably not too harmful—otherwise this large percentage of people would likely all have a similar disease or have died before we got a chance to sequence them. As we collect more and more data, the do-not-disturb list will be fleshed out to include many of the genes that are "embryonic lethal," meaning that you will not be born at all if you have them, and those that are "disease associated," which may be tolerated in some cases.

Once this map is complete, we can start to think about what paths may be altered to ameliorate diseases and help patients. But just as the first real human genome map was finalized in 2021, the discovery of other organisms' genomes has only just begun. The big questions become: How well will our genome do when we get to Mars? What else should we modify when we get there? Can we survive? And what can we create next?

4

PHASE 2: PRELIMINARY ENGINEERING OF GENOMES (2021–2040)

Calling something junk just because we don't understand what it does strikes me as narrow-minded. I suggest replacing this designation with "funk"—functionally unknown DNA. I would prefer to think of our genome as funky rather than junky.

—Dr. Gregory Petsko

ENGINEERING CELLS

By the mid-2030s or 2040s, ideally, we will be able to get human boots on Mars, enabling us to directly see how humans respond to Martian living and how well our "molecular risk mitigation" plans work. Once there, we will be able to test more genetic-engineering designs on more cell types and organisms across a large range of newly defined contexts. As is currently done on Earth (for safety concerns), most of the work will start in animal models and then slowly be expanded to humans. Some possibilities include altering the expression of DNA damage-repair genes, tumor suppressor pathways, or pathways related to cellular and oxidative stress. It will also fundamentally challenge what we think about in terms of a "normal" genome as we continue to make alterations through the selective addition or removal of pieces of the human genome, and we begin to see how many alterations can be made to a cell while still retaining its innate, functional properties. Our current idea of normality will be reexamined from many perspectives—from the human genome to human birth. Pilot experiments for "genetic armor" will also begin, and lessons from two species have already shown this is possible.

LESSONS FROM ELEPHANTS

The first lesson comes from a peculiar genetic fact about elephants. While they are obviously far more massive than humans, with about 1,000 trillion cells to our ~70 trillion, one could imagine that elephants have more chances for something to "go wrong" and to accumulate cancer-causing mutations. But, surprisingly, this is not the case. In fact, they have about a threefold to fivefold lower risk of cancer than humans. The same is true for their even larger aquatic friends—whales. This trend was noticed by many researchers, including Richard Peto, a statistical epidemiologist at the University of Oxford, who was puzzled by this paradox in 1975. It is now appropriately called Peto's paradox. He was first to compare humans to mice, noting that we live about thirty times longer (75 years vs. 2.5 years) and have >1,000 times as many cells. In theory, we have over a million more chances to "get" cancer than mice have, and elephants have >100 times more cells than humans, but they get less cancer. Why?

Subsequent work showed that some of the paradox can be explained by the fact that cells grow and divide at different rates across species, which also changes their risk for cancer. This is true even within a species and a single human body, where different cell types reproduce at different rates. For example, there is a higher incidence of cancer derived from highly proliferative cells such as hemaotoetic cells within your bone marrow, compared to the more stable cardiomyocytes in your heart. But as research progressed, some other peculiarities were noted. First, elephants have a key difference in a gene called TP53. This gene is one of the most commonly mutated genes in cancer, underscoring its importance, and is often called the "guardian of the genome" because it detects DNA damage and subsequently forces the cell to kill itself through a process called apoptosis. Clearly, the gene is important in both humans and elephants, but two 2016 papers from two labs (Joshua Schiffman and Vincent Lynch) gave a clue as to what makes it unique in elephants.

Schiffman's and Lynch's research showed that elephants do not have just one copy of TP53; they actually have *twenty* copies of this gene. The elephants also produce extra copies of the TP53 protein (called p53), meaning the elephant cells are more actively scanning for DNA damage. Indeed, the research showed that elephants' cells are more sensitive to DNA damage from radiation. Their rates of apoptosis (cell death) are much higher than in human cells. Humans only have two copies, as with most genes (one from the mother and one from the father), but the elephants' multiple copies raised an obvious question. Could you just add extra copies of TP53 to human cells and get them to be more resistant to cancer? So far, the answer is yes—though its application will likely be more complex than simply adding more of the exact same gene. By moving a version of the elephant gene into mouse cells and detecting a change in response to radiation that shows activation of apoptosis, called caspase-3, Schiffman and Lynch showed that response to radiation was stronger with the extra gene.

However, a balance of biochemical ingredients and dosage is essential, both in late-night cocktails and human cells' gene expression levels. Indeed, based on current tests of inserting new genes into human cells or altering the expression of current cells, there is a risk of *too much* of a good thing. Any time you tinker with a genome, the tinkering has to be balanced, to ensure the dosage (activity) of the gene is at the correct level.

Such careful genetic dosing is already done naturally within human cells. For example, all females have two X chromosomes in their cells' nuclei, but males only have one. The Y chromosome in males does not make up the difference, harboring only ~200 genes versus the >1,000 genes in the X chromosome. Thus, if both chromosomes were active at the same levels across the sexes, the X chromosome for females would produce "too much" activity. To regulate this difference, the extra chromosome is controlled with "dosage compensation." This regulation ensures that most genes are silenced in the two X chromosomes of women to match the single copy of that chromosome's activity in men. However, just deleting an entire chromosome to match the "dose" of gene activity is not the correct way to solve this balance. Indeed, when females are born with only one X chromosome, it leads to a disease called Turner syndrome. As such, gene expression dosage is a carefully controlled process.

In the case of TP53, uncontrolled high levels have further been shown to cause more rapid aging. How, then, are elephants able to grow both old and cancer free? It turns out that not all copies of the elephant's TP53 gene are identical—some are actually retrogenes (a kind of pseudogene). Some act as guardians of more functional forms of p53, enabling p53 to essentially sit quietly and wait for genetic instability and not overreact. Naked mole rats—which have also been shown to have unique cancer-resistant abilities—further have their own unique processing of p53, which has yet to be entirely understood, but it is likely similar to that of elephants. Integrating one species' cells with those from another species' evolutionary history and learning these "genetic lessons" can act as our guide to protect human cells and the cells of all other species which will inevitably accompany us on our journey away from Earth.

LESSONS FROM TARDIGRADES

An entirely different, validated idea of internal genetic armor comes from an animal called a tardigrade (*Ramazzottius varieornatus*), often called the "water bear" because it can be readily found in water and looks like an adorable microscopic bear. It even has a dedicated Twitter handle (@tardigradopedia). They can survive almost anywhere, including being exposed to the vacuum of space, extensive radiation, and desiccation (drying out). This extremophilic capacity has been known since at least the early 1900s, linking desiccation resistance to radiation resistance, but it was always a mystery as to what gave these little creatures their amazing abilities. Once the genome was sequenced in 2015, multiple research groups (including Atsushi Toyoda and Takekazu Kunieda in Japan and Bob Goldstein at the University of North Carolina) jumped in the race to figure out their superpowers.

The tardigrade's high tolerance for X-rays and other radiation was thought to be a by-product of the animal's adaptation to severe dehydration because severe dehydration can destroy or damage most of the molecules in living things. Just as dry skin can crack, break, and even bleed, the same drying out of cells can damage them at the molecular level. DNA, RNA, proteins, and all the essential components in the cellular soup of a living organism can be torn apart and fractured, as if blasted with X-rays, simply by the rapid loss of water.

So, then, how do the tardigrades survive both desiccation and heavy radiation? How do we know which genes are the most responsible? A simple idea is to see what genes become activated during desiccation and to focus on those, which is what Kunieda's team did first. But the gene expression changes from dehydration to rehydration did not reveal many differences, indicating that the tardigrade can enter a dehydrated state without necessarily needing any large gene expression changes. Given that, the team reasoned that a constitutively (constantly) expressed gene would be a better candidate, or what is called a "housekeeping" gene, because this type of gene is always active in a cell to keep its "house" in proper order—like a molecular housekeeper that never takes a break.

Here is where the unique tardigrade genes came into focus. Most of the genes identified to be unique to tardigrades were both constitutively and highly expressed as proteins. They were also active during the embryonic stage as well as in the adult stages. Yet among the dozens of unique proteins made from those tardigrade-specific genes, only one, which they called the damage-suppressor (Dsup) protein, was found to be colocalized with nuclear DNA. This localization was a clear indication that the protein could work to interact with, and maybe protect, DNA.

So, two tests began—first to see if Dsup can truly improve radioresistance and, second, if it could do this while in an entirely new environment—a human cell. This would require, first, the creation of human cells that had the Dsup protein, and then treatment with X-rays, which can cause genetic havoc in two ways. Of note, X-ray energy can either be directly absorbed by the DNA (direct effects) and break the hereditary molecule, or it can act indirectly, inducing Reactive Oxygen

Species (ROSs) from the water molecules that are activated by X-ray energy. Through genetic-engineering methods (described below), the Japanese group made a HEK293 cell line expressing Dsup under the control of the "constitutive CAG promoter," so it would stay active and make the human cells much like those of the tardigrade.

Then the irradiation began. As expected, exposure to hydrogen peroxide (H_2O_2) induced severe fragmentation of most of the DNA (71 percent) in control HEK293 cells. In contrast, DNA fragmentation in Dsup-expressing cells was substantially suppressed to only 18 percent of total DNA; the Dsup protein therefore seemed able to protect DNA from ROSs as well as X-rays. As a "rescue" experiment, the team also tried giving a pretreatment to the cells with the antioxidant N-acetyl-L-cysteine (NAC), which (as expected) substantially suppressed peroxide-induced single-strand breaks (SSBs). But the superpowers could be combined—when using both NAC and Dsup, there was even greater suppression. This was true for protection for the ROSs but, also, for the double-strand breaks (DSBs), which were a 40 percent reduction in breaks. Not only could the Dsup-expressing human cells exhibit less damage from radiation, they also showed improved viability and proliferative (growth) ability after irradiation.

These results were the first time that a tardigrade gene was put into a human cell, used for radioprotection, and it was shown not to inhibit any of the growth, morphology, or basic functional parameters of the cells. But this was one gene, in one cell line, and not across an entire body. It is unclear what would happen to an entire human. The lessons of the exemplar tardigrade experiments based on elephants and tardigrades are exciting, but what would happen if we went farther? How could we do more genes at once? Could the technique be improved?

Work from my own lab at Cornell has shown that 40 percent of reduction in DNA damage can be increased to 80 percent or even 85 percent by improving the integration and regulation of the Dsup gene, as well as modifying other genes in the human genome. This is the crux of what will happen from 2021 to 2040, when all genes from all organisms will become a playground for creating and making new functions in human cells.

But first a primer on how.

HOW TO ENGINEER A NEW CELL

If an organism is a bowl of soup, then genetic "transformation" is a way to move ingredients from one bowl of soup to another. Multiple kinds of genetic integration and transformation are possible, including different ways to guide foreign DNA into a new cell, methods to grow large quantities of DNA through the usage of bacterial "packaging cells," and transfection, which primarily refers to the integration of DNA into eukaryotic cells. Some of these techniques were developed in the 1950s and 1960s, when they were used to move plasmids (mobile genetic elements) between different bacteria (microbial transfections). The first cloning protocols were also developed then, in order to make copies of these plasmids. But to understand how to engineer a cell, first we need to know how to measure what is even present inside a cell's genome.

Newer methods for cloning were developed in the 1970s and 1980s, as scientists began to figure out ways to clone plasmids and amplify these products with the polymerase chain reaction (PCR)—a chemical reaction that makes continual doubling copies of a target. If you target a specific genetic sequence, and you use "primers" at the end and beginning of that sequence to "prime" the reaction, you can then begin to extend a copy from a single piece of DNA. Thus, one copy becomes two, then two becomes four, and four becomes eight, at a rate of 2^n number of copies. This simple method, for which Kary Mullis received the Nobel Prize, ushered in the modern age of molecular biology.

Indeed, PCR was the gasoline for the beginning of the "genome reading" era when scientists everywhere lit a match in their minds and amplified the DNA of anything they desired. This revolutionary technology was key in the mapping of the human genome. These technologies also led to the first bacterial genome, in 1995, by Craig Venter (Haemophilus Influenzae Rd), and then the fruit fly genome, in 1999, by Gerald Rubin, and many others afterward. Once we could sequence the DNA from various organisms, we could begin to see what happened when we performed transfections and genome manipulation at base-level resolution.

Just as some people get tan when they go to the beach on a sunny weekend while others get sunburned, different kinds of bacterial and other cells can be more readily "transformed" than others. Some are very easy to transform—they happily absorb DNA from their environment—whereas others are much more difficult. *Deinococcus radiodurans*, for example, is notorious for being able to absorb bacteria and then immediately put it to use. *E. coli* is another good example of a DNA sponge, an ability that has cemented its place as a mainstay of modern molecular biology.

However, most organisms do not like invasion by non-self DNA—the cellular equivalent of a stranger suddenly putting food in your mouth on the subway. For example, plants have a built-in, hard cellulose cell wall to help ensure no DNA floats inside. Eukaryotic cells, like those of humans, have active enzymes (DNAses) and methods to defend against any invading DNA or RNA (RNAses), although they are not perfect. These enzymes and physical structures are part of an ongoing defense against viruses, bacteria, and other foreign organisms that is perpetuated by the individual cells, as well as the active immune system for fruit flies and humans alike.

Because human cells do not like to take in foreign DNA, it has to be introduced artificially in a process called a "transfection," which sounds a bit like what it is. You need to take an infectious agent and transfer it from one organism to another in order for that host organism to take up that foreign DNA. This can be done with an infectious agent that can sneak directly into the cell through the use of molecules already on the surface or a physical process that opens up the membrane of the cell, allowing the DNA to enter.

VIRAL METHODS FOR DNA CARGO
For both transfection and transformation, there are multiple ways to prime the system so it can accept the new DNA. For transfection, you need to take naked, purified cells and enable them to become more absorbent. This process can include methods such as calcium phosphate coprecipitation, liposomes, electroporation (electricity zapping of the cells, like Frankenstein's lightning), gene guns (quite literally shooting genes into the cells), or microinjection. For transformation, this process is much simpler because most bacteria will happily absorb the DNA, so

you just need to do some electroporation or chemical transformation to prime the cells.

In human cells, there are two kinds of transfection, depending on whether it is temporary or stable. Transient (temporary) transfection means that the new pieces of DNA come and go, usually along with the phenotype (trait) that is being transfected. This allows us to test the impact of the expression of the "transgene" over a short period of time. This nonpermanent change may even be ideal in therapeutic interventions, where the expression is only required for a short duration to accomplish a clinical goal, and prolonged expression may result in increased risk to the patient. Stably transfected cells, as the name implies, are those cells which have permanently integrated the foreign DNA into their genome. If you want to know the continual change of a cell's regulatory processes or how an organism's development across its life may be altered from a piece of foreign DNA, a stable transfection is the best path. It means that the genetic payload is now a component of the host genome, just like the other genes passed down from its parents, permanently integrated. Within this design, many different elements may be added over time to assess how this additive process may affect the overall function of the cell.

However, with the stably transfected cell line, the mechanism of getting the foreign DNA can require the ravaging actions of a virus. A common method is a lentiviral infection or an adeno-associated virus (AAV). In both cases, you need a virus to infect the cells, smuggle the genetic payload into the nucleus, and then integrate into the host genome, like planting a new species of tree in a previously hidden forest. Once this happens, it may create havoc if it inserts near a region that controls essential genes, such as tumor suppressors, and further disrupts the host regulatory framework. This is like planting an alien tree in the middle of a small stream and disrupting the flow of water. Moreover, the integration might not occur in just one place per cell, but it could occur in ten or more locations across the host genome—each unique to every engineered cell. The lack of predictability of this heterogeneous event can create some additional problems of "dosage," where each of the integrated gene's effect is additive, resulting in a greater response with more integrations per cell. Therefore, stable transfection can give you long-term results and a stable phenotype, but it can also create additional chances for things to go awry during the process.

Moreover, not all human cells are the same, as clearly evidenced by the different tissues in your body. This is due to the alteration of epigenetic ("on top" of the genome) regulation as your cells are programmed to differentiate from your original single cell to the person you are today. Epigenetic regulation is the process that controls which part of the 3.1 billion bases of your human genome are "open." Genes in each human cell are activated and utilized, depending on whether they are primed for use, in a DNA-protein structure called chromatin (the packing of the DNA). If the chromatin is open, it could be a prime site for a virus to integrate and "hop in."

While this sounds creepy or even intrusive, it is actually surprisingly common throughout human evolution and our own lives. Look no further than your own cells to find ample evidence of viral integration. If your genome is a 100-page book of life's instructions, 8 percent of that book's pages are viruses. There are many kinds of endogenous viruses in our genome, as well as those in the genomes of other mammals and plants. Transposons are "jumping genes" that can "copy and paste" fragments of DNA within the same genome. Though these were first discovered by Barbara McClintock to explain the variegated colors of corn kernels (and led to her Nobel Prize), work by Dr. Alex Kentsis and our group have also shown that aberrant transposons can lead to cancer. Retrotransposons, like transposons, also move by a "copy and paste" mechanism, but they hop around using RNA as their copying substrate instead of DNA. Similarly, human endogenous retroviruses (HERVs) are viruses in the human genome that can activate, convert from RNA to DNA and then integrate back into the genome. A well-known virus that also integrates into the host genome is HIV, but this event is just the tip of the iceberg of "genetic tagging."

Beyond HIV, other viruses can also dive right into your genome and make themselves at home, like a bad college roommate who makes a cellular mess, eats your food, and doesn't pay rent. One of these is human papillomavirus (HPV), of which certain strains are responsible for >60 percent of cervical cancers. These viruses integrate into the host genome and disrupt gene expression. Herpes also integrates into the genome and can make a long-term home in your cells, which led to the birth of the phrase, "Love is fleeting, but herpes is forever." These integrating viruses replicate in the nucleus, whereas other viruses replicate in the cytoplasm (e.g., Zika, West Nile Virus, and HCV) but do not integrate into the genome.

Humans are not alone in the ongoing battle with viruses. Pigs have their own endogenous retroviruses called porcine endogenous retroviruses (PERVs), which are one of the largest challenges in synthetic biology. A company called Editas Medicine is working to engineer the PERVs out of the pig genome so that human organs can be grown inside pigs and then transplanted to humans to ameliorate the human-organ shortage. If the PERVs were not stopped before transplantation, the organ would be rejected by the human's immune system (more in chapter 5).

But! Viruses can make a home in human cells even without direct genome integration. For example, most people get chicken pox when they are young, and then they resolve the infection. Yet the chicken pox virus, like many other surreptitious infectious agents, can have a mind of its own and stay around inside human cells. The virus does not get removed from the body, but rather it hides inside the neurons in various parts of the body and can then activate later in life as shingles, which occurs when the immune system is suppressed and/or the virus reactivates to create a painful linear infectious rash, usually in older people.

Finally, some cancers can even be contagious like the flu. Tasmanian devils look strange—they appear to be creatures created from a bad dream about rabid dogs having sex with shy bats. Indeed, when they mate, they often bite each other. But even before the mating, their form of conception involves a lot of growling and face biting as foreplay (no offense if you're into that). This process enabled the propagation and sustenance of a cancer-causing virus being passed from face to face of the Tasmanian devils for decades. Estimates are that a single clone of these virus-infected tumor cells have been around for at least a hundred years and caused the facial tumors for many of these devils.

Given this context, the idea of using viruses as a genetic eighteen-wheeler transmission vehicle for delivery may still seem unsettling, but it is worth recognizing that it common across biology. Fortunately, the safety profile and understanding of these methods have already drastically improved since their inception. They are currently used in many different therapies—and will likely continue to be used throughout the 2040 timeline—while methods of genetic payload delivery, editing, and monitoring are continuing to emerge which decrease their risk while enhancing their predictability. It is only a matter of time for the risk of genetic engineering to be mitigated entirely, and we already know how it can be done.

ENGINEERING GENOMES

THE OLD METHODS OF GENE EDITING

Since the first recombinant DNA technologies of the 1970s, the unpredictable nature of inserting new DNA into an organism's genome led to uncertainty about their safety and applicability for the clinic. However, eventually the promise of the technology was too lucrative, given that it was the clearest way to replace a broken gene with a functional copy. As soon as DNA could be readily modified, it was only a question of how, not whether, to use gene therapies.

One of the first big successes for gene therapy came in 1990, for Ashanthi DeSilva, a young patient with "bubble boy" syndrome, from severe combined immunodeficiency, or SCID. This disease is severe; it prevented her from having an adaptive immune system and, thus, provided no way to fight off a pathogen or bacteria. To fix the SCID, scientists and doctors drew her bone marrow and purified white blood cells (WBCs), and then used a retrovirus to insert a working gene into those cells. They then injected the WBCs back into her body, which helped to give her a functioning immune system, and she could finally go outside with freedom at age five—for the first time since she was born.

French researchers conducted a similar trial on SCID with ten children in 2002, using stem cells taken from the patients' bone marrow in order to make functional WBCs to fight off infections. Excitingly, the treatments worked, and the SCID seemed to go away. But, by 2007, four of the children developed leukemia, and one of them died. Here, again, the culprit was the randomness of the process of retroviral integration.

For at least one of the children, the virus had been integrated near an oncogene, which can lead to leukemia or other cancers. But these are not the only cautionary tales.

The most notorious case of gene-therapy failure was that of Jesse Gelsinger, who used an adenovirus vector to treat a disease called ornithine transcarbamylase deficiency syndrome (OTCD) in which ammonia builds up to lethal levels in the blood. Jesse was relatively healthy, and he could control his OTCD by being very selective about the foods he ate, which could keep the ammonia levels in his blood down. Even though Jesse was the eighteenth person to get the therapy in the clinical trial, he quickly showed flulike symptoms, discomfort, and signs of jaundice. He had an intense and painful inflammatory response, followed by kidney, liver, and lung failure, and he had to be put into a coma to survive. Within four days of the therapy's initiation, he was dead.

Jesse's death in 1999 led to a global scare about genetic engineering and gene therapies for any disease as well as comprehensive investigations into the possible causes. An investigation found that two patients in the trial suffered serious side effects and that the scientists did not immediately inform the agency or put the study on hold as required. Also, Jesse's blood work performed before the gene therapy showed that he had poor liver function, which meant he was not likely a good candidate for the study after all. Indeed, research from the US Food and Drug Administration (FDA) and National Institutes of Health (NIH) showed that 691 volunteers in gene-therapy experiments had either died or fallen ill in the seven years preceding Jesse's death; only 39 of these incidents had been reported promptly.

James Wilson, who was the director of the University of Pennsylvania's Institute for Human Gene Therapy—and who led the trial where Gelsinger died—desperately wanted to understand what happened. He looked into using other means to safely conduct gene editing therapies, finding several candidate (adenoviruses and AAV variants). Nonetheless, the FDA charged Wilson with several clinical-trial violations, and in 2005, he agreed to restrictions on his human research for five years, with the university agreeing to a $514,000 settlement to the government. The Institute for Human Gene Therapy eventually closed down.

The Gelsinger death, the subsequent scare, and related regulatory changes led to a slight decline in new genetic-engineering clinical trials in 2000–2001. For a while, in the early 2000s, even saying you were working on "gene therapy" was a professional liability in some scientific circles. But the promise of using gene therapies was too big to resist. The field was awash with ideas and concepts on how to improve the treatments, the delivery, and the packaging of the genetic payloads.

Scientists and clinical staff could not help but ask: Can we make the treatments safer? Can gene therapies work as well as we hope? Scientists were publishing a variety of new gene-therapy animal models, building on new ideas of gene engineering, and there were multiple paths ahead to make them safer. Even in human cohorts, the clinical trials using genetic-engineering methods continued to grow.

While the media covered the Gelsinger death as a challenge to the whole field of therapeutic gene editing, the purported scare over trials did not actually lead to a cessation of starting new ones. In fact, in 2000–2001 alone, almost 200 new genetic-engineering clinical trials were created. The scary newspaper headlines of the early 2000s belied the clinical reality, because clinical trials using genetic-engineering

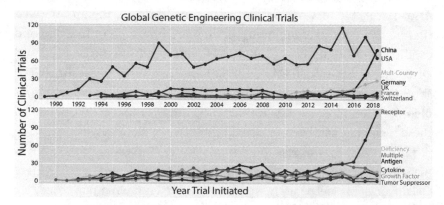

4.1 The rise of genetic-engineering trials. Top: The number of trials initiated each year within a given country; only the seven countries with the most trials are shown. The United States had the most number of trials every year until the most recent year analyzed, 2018, when they were surpassed by China. Bottom: Types of engineering treatments continuously emerged, with an especially recent rise in receptor-based therapies largely due to chimeric antigen receptor t-cell (CAR-T) therapies targeting cancer.

methods continued to steadily climb, especially for cancer, from the late 1990s until the 2020s and onward.

The interest was (and still is) clearly present; all we needed were more accurate methods to deploy such genetic engineering. The random insertion of a new genetic payload was always viewed as risky. We needed new methods for genome editing and modifications, which came quickly.

THE NEW GENE-EDITING METHODS

Revolutionary changes have occurred since the early gene-therapy trials. There are now multiple methods for precise editing of specific genes, superseding the methods that were always dependent upon the semirandom addition of new genetic material from adenoviruses and AAVs, which likely led to the issues observed with Gelsinger. There are further improved ways to enable quality control of these methods, so that if methods with random integration are employed, they can be checked before therapeutic delivery. These new methods of genetic engineering began in the early 1980s, and fall into several categories around both enzymes and genetic constructs.

The first tools of genome editing came in the late 1980s, from enzymes called meganucleases, which are part of a family of enzymes called endonucleases. As the name indicates ("endo," inside, "nucle," nucleus of the cell where DNA lives, and "ase," enzyme), these enzymes can match and cut pieces of DNA. A key first step in genome editing is to actually break the DNA (double strand break, or "DSB") at a selected site of interest. Then, you can insert the new sequence you designed, repair and patch up the DNA, and send the cells on their way. It is akin to breaking your own legs to make yourself taller. They create DSBs at very specific sites, ranging from 14 to 40 nucleotides, and thus they are very specific to a particular spot in the genome. But there is not a meganuclease for every known sequence of DNA, and this lack of modulation largely limits their therapeutic potential.

Thus, newer methods were needed that could be more scalable, and indeed three of them have emerged since the 1980s for genetic engineering. These include (1) zinc-finger nucleases (ZFNs), (2) transcription activator-like effector nucleases (TALE nucleases, or TALENs), and (3) clustered regularly interspaced short palindromic repeats (CRISPR). The first clinical trials using these gene-editing methods appeared in 2008, and the number and type of these trials has continually increased since those early trials to now include all three of these methods. They have progressed so quickly since 2015 that they are now standard teaching for high school and college science classrooms.

Zinc-finger nucleases The research behind ZFNs leveraged knowledge gained from research on zinc-finger transcription factors (TFs). TFs are master regulators of gene expression, constantly scanning the genome to see which genes should be turned on and which should be turned off. If your genes are appliances in your kitchen (refrigerator, toaster, stove, grinder), then the TFs are the hands reaching to control the "on" and "off" buttons for your cell's appliances. Because these enzymes have already been shaped by hundreds of millions of years of evolution to read specific DNA sequences (called motifs), they can deftly control your cell's functions when they find the specific motif they recognize, just like quickly recognizing the face of someone you love. Thus, ZFNs become obvious candidates for enzymes that could be used to exquisitely target sequences of DNA.

However, even TFs have their own regulatory levers and switches. For example, when "hugging" a zinc ion, ZFN TFs become more stabilized, enabling their ability to recognize their given motif and interact with the DNA. Thus, if a diseased genome is an open page of text on which you want to run a "find and replace" function, then the TF is the "find" function, the zinc ion is the "enter" key, and all that is needed is to pair it with another enzyme for the "replace" function.

ZFNs perform the "find and replace" by merging pieces of two proteins together (called a fusion protein). Most proteins have multiple regions (domains) in their large, three-dimensional structures, with subsections of these 3D structure that can be merged together, and ZFNs are no different. The part of the nuclease that actually cuts the DNA (like a meganuclease) can therefore be linked to protein fragments that only recognize specific sites (like ZFN TF motifs), which then enable cutting only at specific sites across the genome.

However, each single TF motif is very short (three or four base pairs) and, thus, not very specific. In a genome with four bases, using three bases to generate a diversity of targets would only include 4^3 (n = 64) or 4^4 (n = 256) combinations, which is far too limiting for enabling targets of multiple types. To get around this challenge, researchers in the Giordano laboratory in 2002 used seven to eight of these ZF recognition sites together in a single construct, so that the ZFN could target twenty to twenty-four bases at one time. This enabled the team to target genes required for angiogenesis in a mouse model of cancer in one of the first demonstrations of the specificity and power of ZFNs. In 2008, they were also used to target glioblastoma cells (in vitro) and increase their susceptibility to treatment (glucocorticoids). But this is not the only way to edit a genome.

TALENs TALENs are similar to ZFNs, in that they also merge the DNA-cutting protein domain of a nuclease, but they use a "homing beacon" site-recognition power of a TALE instead of the ZFN TF. However, their recognition domain is much longer, based on a highly conserved sequence of thirty-four amino acids. These proteins are excreted by a plant pathogenic bacteria, called *Xanthomanos*, as a tool to reprogram their plant hosts by binding to their promoter sequences in the host plant and controlling the activation of various plant genes that help the *Xanthomanos* during infection.

Since TALENs have a longer "recognition domain" than those of ZFNs, they quickly became the preferred method of genome editing in the early 2010s. The long (>30 base pairs) binding sites from the TALENs gave them greater specificity and lower chances of an "off-target" cutting in a site that one would not want to edit (such as oncogenes), and also their regular sequences made them easier to construct. All these features together meant that TALENs arguably could give the most precise editing of the most appropriate site to treat a disease, until CRISPR emerged.

CRISPR CRISPR is similar to both TALENs and ZFNs in that it, too, has a molecule that helps guide it along the genome, identifies a site of interest, cleaves the DNA, and then incorporates the wanted changes. However, the revolutionary feature of CRISPR is the ease with which genome editing occurs. CRISPR works by using a guide RNA (gRNA), which quite literally guides the machinery into place resulting in highly specific DNA breaks. The important component here is that the "scanning element" of this technology is an RNA sequence—not protein motif—which allows for the DNA recognition. This makes it possible to directly and easily synthesize the molecule necessary to target specific regions of the genome as soon as you know the sequence. Given the complexity of converting nucleic sequences into the resulting three-dimensional shape of their amino acids, it is much more straightforward to synthesize novel targets with RNA than synthesizing novel proteins with highly specific motifs.

THE CRISPR REVOLUTION

The history of CRISPR, as well as its tools, methods, and underlying mechanisms, provides an understanding and appreciation for how an accidental finding can rapidly develop into a revolutionary medical tool. Thanks to decades of human innovation and engineering, CRISPR is now an integral part of molecular biology and clinical care. This rapid development of biological editing and engineering methods will continue to accelerate in the coming 500 years. Also, it will likely continually replay an auto-enriching dance between comparative genomics, careful mechanistic experiments, and simple yet powerful observations of biology. A quick review of the background on how CRISPR was developed is essentially a preview of what we might find as we explore other planets.

FINDING CRISPR

Like many things in science, CRISPR was discovered by accident. The method was first found in 1987 by Japanese researchers (including Yoshizumi Ishino) when they were cloning a part of an enzyme they were studying called *iap* (*isozyme conversion of alkaline phosphatase*). As part of the cloning process, they noticed unusual repeats present in their cloned DNA. In genetics, repeats are often in consecutive segments, or in tandem, but these repeats were very different.

At the time of the Japanese group's discovery, it was not clear what the repeats meant, but more evidence came quickly. In 1993, a scientist named Jan van Embden and colleagues in the Netherlands were studying *Mycobacterium tuberculosis*, the organism causing tuberculosis, and they, too, noticed something strange about the bacterium. It had clusters of repeated sequences, which they called "interrupted direct repeats," and these showed a diversity of sequences across different strains of tuberculosis. At first, van Embden's team, too, was puzzled by the repeats; but they quickly realized that the divergence of these repeats was strain specific, so they could make primers that could exquisitely target one strain of *Mycobacterium tuberculosis*, such as one that is more virulent versus a benign strain. They could use oligonucleotides (primers) for each species to quickly genotype the strains that were present, in a process, called spoligotyping, that is still used today. As more and more sequence data became available in the 1990s, it was possible to look for these trends in other microbes too. Bacteria, archaea, and anything that could be sequenced were now being profiled with the Sanger automated sequencing method, which led to a boon in data for pure discovery, and this included anything strange.

The plethora of DNA-sequencing data available led to the birth of the fields of *bioinformatics* and *comparative genomics*, which blends computational modeling, information analysis, sequence analysis, and computer programming with biology. One of the first bioinformatics scientists was Francisco Mojica, a graduate student at the University of Alicante in Spain, where salt marshes held curious organisms that could survive extreme salinity. Essentially, these are like looking for strange organisms on an alien planet. For example, *Haloferax mediterranei* is an archaeal microbe with extreme salt tolerance that had been isolated from Santa Pola's marshes in Spain, and these were being studied by Mojica. His advisor had found that the salt concentration of the growth medium changed how well the restriction enzymes cut the microbe's genome, and Mojica's project was to figure out why.

When he got the sequence data back, Mojica also noticed strange repeats. He saw multiple copies of a palindromic (meaning it is the same forward and backward), repeated sequence of thirty bases, separated by "spacers" of roughly thirty-six bases—that did not match anything he had seen before. He published these results in 1993, citing the 1987 work by the Ishino group, but it was still left as mostly a mystery, and he continued to look at more and more sequences to figure out why.

In 1999, Mojica started his own laboratory at the University of Alicante, and the first thing he did was to begin scanning large databases of archaea to look for patterns of repeats, mostly still in *Haloferax* and *Haloarcula* species, but he noticed the same patterns consistently showing up in various other species. This led to a comprehensive scan of new species as they emerged in the literature, and by 2000 he had found these repeats elements in twenty different microbes, a finding indicating that there was a strong evolutionary reason to keep these kinds of repeats present in the DNA of organisms from across the world. Also in 2000, Mojica noted that transcription (turning into RNA) of the interrupted repeats was also happening in the cells, meaning that these sequences were activated in the cells (not just sitting as DNA).

In 2001, Mojica and Ruud Jansen, who were both searching for additional interrupted repeats, proposed the acronym *CRISPR* to merge the phalanx of names appearing in the literature. It stuck and was quickly adopted by other researchers. Another feature of the CRISPR elements was that in prokaryotes, the repeat cluster seemed almost always to be accompanied by a set of unique genes called "CRISPR-associated systems," or Cas genes. From work by Jansen and Mojica, four Cas genes (Cas 1–4) were initially found and described. They were then examined at the protein level, where they showed helicase and nuclease motifs, which meant they could potentially unwind DNA and also cut DNA. However, beyond these early hypotheses, the CRISPR function remained enigmatic. The true purpose of these repeats was not clear.

THE PURPOSE OF CRISPR

In one of the most powerful examples of the power of computational biology and bioinformatics, it was a computer algorithm and hard work that gave the biggest clue. Mojica spent most of the summer of 2003 using an alignment program called BLAST (Basic Local Alignment Search Tool), comparing the CRISPR repeats he observed to any other sequences that existed. Even though he had done this dozens of times before, this process was worth repeating as often as possible because DNA databases are constantly updated and ever-expanding. Mojica got lucky when he found a spacer that matched a phage (a virus that infects bacteria) called P1, which can infect the *E. coli* bacteria. This immediately connected an adaptive genetic system in the bacteria (CRISPR array) with the precise sequences from the viruses that had infected them (the phages), and thus, a new defense system inside bacteria was found. It turned out that all those bacteria Mojica had been studying had these CRISPR systems as a *primordial bacterial immune system* to remember any viruses that had infected them.

Other researchers with access to different databases quickly confirmed these results, including a team from the French Ministry of Defense (including Gilles Vergnaud) and Alexander Bolotin, a Russian microbiologist at the French National Institute for Agricultural Research. These additional validations in *Yersinia pestis* (plague) and other bacteria confirmed this mapping of phages and their targets, as well as the adaptive nature of the system. By 2003, there was effectively an entirely new CRISPR field. In the future, the exact same method of comparing sequences found on other planets can help reveal how biology is adapting, and simultaneously provide clues as to how it works.

It didn't take long for research groups around the world to start teasing out the potential of CRISPR. The first experimental evidence (rather than just comparing sequences and inferring it) that CRISPR was a "bacterial immune system" came in 2006 from Rodolphe Barrangou. Then, the first experiments to reprogram CRISPR came in 2008, from Luciano Marraffini at the University of Chicago and Erik Sontheimer at Northwestern University. They had both been working hard to decipher the exact target of the CRISPR system (e.g., RNA or DNA) and ways to build it from scratch.

UNRAVELING THE MECHANISMS OF CRISPR

However, it was not yet clear how exactly it worked inside the cells. From 2007 to 2008, two researchers (Moineau and Danisco) focused on various bacteria where CRISPR did not work as well and where plasmids could be only partially defended against. They confirmed that the cutting of the plasmid depended on a Cas enzyme (in this case, the Cas9 nuclease). But here again, they sequenced the products and looked at the sequences to figure out why. After examining the data, they found a set of three bases near the cutting site, which they called the protospacer adjacent motif (PAM) sequence. They showed that viral DNA is also cut in precisely the same position relative to the PAM sequence, meaning that this was part of the "homing beacon" for the bacteria to cut at a specific site. Even more powerful was the fact that as more spacers matched the plasmid, the number of cuts also increased. It was a titrated, targeted system.

The second key mechanistic part came from John van der Oost and Eugene Koonin, who found that they could move the entire CRISPR system from one bacteria into another, effectively "rebooting" the function and also reprogramming it. They found different kinds of CRISPR systems in various bacteria (Class 1 vs. 2), which they noted all had different sets of Cas enzymes. But for all these bacteria, these sets of coordinated enzymes were all activated into one large framework (they called Cascade), which is needed to turn a pre-RNA into a more processed version of the RNA (sixty-one bases) called the CRISPR RNA (crRNA). Van der Oost and Koonin noticed that the last eight bases of the repeat were followed by the spacer, and then the beginning of the next repeat, making the RNA fold into a functional structure that enabled the targeted homing and cutting. They used this design to make a synthesized version of the first-ever artificial CRISPR array— effectively a customized vaccine that could potentially be designed for any bacteria.

Two other researchers (Marraffini and Sontheimer) next planned to re-create the entire CRISPR system in vitro, but it was too difficult in their bacterium of choice (*S. epidermidis*), since it possessed nine Cas genes and was going to take a long time to characterize. Instead, Marraffini and Sontheimer modified the plasmid being targeted by the *S. epidermis* CRISPR system. They added a "self-splicing" element that would not function if the *S. epidermis* CRISPR system was working on RNA as a substrate, but it would work on DNA because the insertion would mean the CRISPR spacer would no longer match, and the immunity and function would be lost. Their results showed that CRISPR worked on DNA, not RNA, and that it was effectively a "programmable restriction enzyme," as they and Eric Lander stated.

Marraffini and Sontheimer were then the first people to declare that CRISPR might be repurposed for genome editing in other cells, including those of humans. They subtly noted in their research paper a gigantic understatement, "From a practical standpoint, the ability to direct the specific addressable destruction of DNA that contains any given 24- to 48-nucleotide target sequence could have considerable functional utility, especially if the system can function outside of its native bacterial or archaeal context."

The final piece of the mechanistic puzzle came from Emmanuelle Charpentier and Jörg Vogel in 2011. Charpentier had been looking for microbial RNAs that could have function, and she met Vogel at a meeting in Wisconsin, surrounded by excessive cheese, beer, and Midwestern wholesomeness. Vogel had recently used high-throughput sequencing (next-generation sequencing, or NGS) methods to better understand the RNA of *Heliobacter pylori*, which can lead to stomach ulcers. This method of "shotgun" sequencing does just what it sounds like—it breaks apart all the fragments of RNA or DNA and sequences them, which can then be mapped to the host genome's DNA sequences in databases—just as Mojica did with the BLAST algorithm in the Spanish summer of 2003, what is done with the subway bacterial DNA in the MetaSUB Consortium, and what we do in the Mason lab every day.

When Charpentier and Vogel looked at the RNA from Charpentier's bacteria of interest, which she had been studying for years (*Streptococcus pyogenes*), they noticed something very strange. They found a highly abundant, yet small (<100 nucleotides) RNA that had a near-perfect match to the CRISPR sequences and was the third most abundant type of RNA in the cell. The only other RNAs more abundant were those that make proteins (ribosomal RNAs, rRNAs) and those that mediate the transfer of information to make proteins (transfer RNAs, or tRNAs). This was shocking—how could something so abundant have been missed all this time? It was simply the size, and the ability to sequence it, that had become much easier by 2011, and made such a discovery possible. Charpentier and Vogel called it trans-activating CRISPR RNA (tracrRNA). They confirmed that this tracrRNA was essential to have the CRISPR system function, and it served as the last needed piece of the homing beacon.

CRISPR FOR HUMANS

Then, thanks to the work of Jennifer Doudna and Virginijus Šikšnys, CRISPR became a revolutionary tool, not just a curious feature of bacterial immunity. Charpentier met Doudna in 2011, and they began to collaborate on ways to make a simpler system for editing. They demonstrated in a completely artificial system (in vitro) (1) that the Cascade system could cut DNA, (2) that it was possible to use customized crRNAs, and (3) that both crRNA and tracrRNA were required for Cas9 to function. But importantly, they showed that all these mechanisms could work just as well when fused into a single-guide RNA (sgRNA), which was complementary to the target of interest. This meant you could begin to view the genome as a document to be edited, and evolution had a new toolkit, being driven by the human mind and imagination.

Once the mechanism was solved and showed that it *should* work in other organisms, it still remained unclear that it *could* work in mammalian cells. From 2012 to 2013, a bonanza of experiments testing these ideas showed that, indeed, CRISPR functioned in mammalian cells. Feng Zhang and George Church began to plan ways to test out the system in human cells. But to do this, he first needed to make a version of the Cas9 enzyme that was "codon optimized" to work in human cells. This is both a key feature of genome engineering, as well as a key feature of genome design, on this planet or others.

Codon optimization is the process where a protein sequence from one organism (e.g., bacteria) is prepared for utility and expression in another organism (e.g., human) by matching how often codons are actually used inside the cell of that second organism. Codons are the intermediary in the readout of the central dogma of molecular biology (DNA to RNA to protein) that build many of the functional parts of cells. Since there are four letters of the genetic code and codons are three-bases long, there are $4^3 = 64$ codons. These sixty-four codons are used by almost all organisms on Earth, including one "start" codon and three "stop" codons for the genetic code. This means that there is some redundancy in the genetic code, with sixty codons to create matches for twenty amino acids, which use the tRNAs described above to "match" one amino acid to one triplet of the genetic code.

But since there is some redundancy, and organisms adapt and evolve in different environments, their abundance and frequency of use is not the same across different species. For example, valine is an amino acid encoded by four different codons (GUG, GUU, GUC, and GUA). In human cell lines, the GUG codon is preferentially used over the GUU and other codons (47 percent use vs. 18 percent, 24 percent, and 11 percent), but this is not true for *E. coli*. It has a different codon preference inside its cells because it uses the same codons at the rate of 35 percent, versus 28 percent, 20 percent, and 17 percent, respectively. Thus, to design a protein to work in one species or another, the choice of codons that are in your protein should be optimized to match the normal use of those codons (and thus, amino acids) in the cells of that species. Codon usage has been mapped out for any species with a reference genome today. (The GenScript database hosts all these.) Mapping is easy to do today, but it was almost impossible until enough genome sequences were available to use.

Once the construct was codon optimized, it then needed to be controlled in a mammalian cell. Zhang then added a "nuclear localization" signal to his Cas9 enzyme, which means it would move to the nucleus of human cells and enable the cutting. But even then, the function was modest—the cutting and editing did not work as well or as efficiently as Zhang had hoped. He tested different Cas enzymes from different species and found that the one in *Streptococcus pyogenes* worked much better. Also, even though human cells did not have the array of other bacterial enzymes that could process the RNA (e.g., bacterial RNase III), human cells could still process the crRNAs and function, and the right sequence of tracrRNA was also found by Zhang. By 2012, he showed that sixteen sites could be edited at once in human and mouse cells, and then he read the sgRNA work from Charpentier and Doudna, which made the system even simpler. He updated the system to show that an updated sgRNA (which solved part of the structure of the RNA) worked very well and then could be used in mammalian systems. Indeed, Church and Zhang showed that full-length fusions of the crRNA-tracrRNA fusions could work well in human cells, and they were finally able to open the toolbox of bacteria to potentially any human or mammalian cell.

OPTIMIZING CRISPR

The race was then on to perfect this system. In 2013, Doudna and Church collaborated to show precise editing at one site in the human genome. Dozens of other groups began to use and test these systems, leveraging the nonprofit site called AddGene, which is a repository of constructs, cells, and genome-editing tools and protocols. Then Jin-Soo Kim from Korea showed that the full-length fusion of crRNA-tracrRNA could be used to make germline changes in zebra fish. This meant that mammals, vertebrates, and potentially any other system could be edited at will. For their work in creating this system that has revolutionized genetics, Drs. Doudna and Charpentier won the Nobel Prize in Chemistry in October 2020.

However, in 2018, a new problem emerged after two different groups noted unintended consequences of the CRISPR editing. Since the CRISPR enzymes are essentially scissors that cut DNA across both strands, the DNA needs to be repaired—and the cell took notice. Our old friend TP53 (as discussed with the elephants) saw this damage and was activated to help repair and clean it up. It began scanning the genome, sensing that something had gone wrong.

Granted, this is a normal part of a cell's life. The p53 gene also gets activated when cells are damaged by radiation. Because of this, p53 is a very important gene that helps protect against DNA damage. To stop cells from becoming cancerous, p53 can help guide a cell to self-destruct, in a process called apoptosis, if (or when) a cell no longer responds to DNA damage sensors. But if p53 is mutated, this safety mechanism can be broken. Indeed, in cancer, when p53 is mutated, it can lead to a sudden emergence and rapid growth of the mutated cell. It is especially pronounced in ovarian cancer, where 95 percent of tumors show mutations in this gene. Hence, the challenge: if you have a batch of cells, some with wild-type (non-mutated) p53 and some with mutated p53, and you damage the DNA with CRISPR, the cells with less capacity to self-destruct will have a better chance to repair and continue to live to see another day, while their wild-type brethren all self-destruct to save the rest of the organism.

Sadly, this is exactly what was happening in the CRISPR cells in therapy. It turned out that CRISPR worked better in cells with a dysfunctional p53 gene. CRISPR had stopped one of the body's key disease-fighting mechanisms, making healthy cells die and enriching for potentially cancerous cells. It was basically an evil form of evolution and selection pressure. Something akin to fixing your melanoma but leaving scars all over your skin. As George Church once said, it is basically "genome vandalism."

But in 2019, a significant update to genome editing came with the introduction of "prime editing." David Liu and colleagues at the Broad Institute of MIT and Harvard finally made a more precise version, by using an impaired version of the Cas9 endonuclease and a prime-editing guide RNA (pegRNA). He and his team had a goal of optimizing the CRISPR system to be more precise, incur less off-target effects, and avoid the problems caused from DSBs. Liu altered Cas9 so that it only makes a cut on one strand of the double helix, rather than both strands, effectively avoiding the p53 selection problem. The entire optimized CRISPR machinery anchors on to the target site and carries the desired edit as well as a new piece of equipment—a reverse transcriptase (RT). The RT enzyme can convert RNA to DNA, which then uses the "fix me" information embedded in the pegRNA to patch up the site of interest.

This exciting new system was tested already in a variety of cell types. Liu corrected a single-base error that can cause sickle cell anemia (called a transversion, in the HBB gene), changed four co-existing variants (a deletion in HEXA) that can lead to Tay-Sachs disease, installed a "protective transversion" in PRNP, and inserted tags and epitopes into target loci. Overall, they performed 175 edits in human cell lines and primary postmitotic (postdividing) mouse cortical neurons.

The big excitement was that prime editing showed a higher or similar efficiency for creating the changes in the genome and had fewer byproducts than did homology-directed repair. Specifically, prime editing induced a much lower off-target editing rate than did the normal Cas9 nuclease at known Cas9 off-target sites (10 percent vs. 90

percent off-target effects). The headlines were exciting—prime editing expanded the scope and capabilities of genome editing, and in principle could correct up to 89 percent of known genetic variants associated with human diseases, based on the mutations that could be addressed with pegRNAs.

But here, too, the system is not perfect, as noted by Jonathan Wilde and other researchers in the field. The pegRNA methods were tested in vitro in a human system, not in an actual body, which is obviously more complicated. Further, ramping up an RT enzyme inside a cell might work in a dish, but within a human body, the immune system might not respond well leading to their targeted destruction. Finally, activating and reverse transcribing (DNA into RNA) rampantly in cells is exactly what can lead to issues with retroviruses, as described above. So, while the excitement continues to grow, the continued hunt for new, better genome-editing methods is ongoing.

HUNTING FOR, AND FINDING, MORE TOOLS

Indeed, from 2003 until today, there is an ongoing search to scan all known bacterial genomes to find new kinds of CRISPR arrays and new types of editing constructs. In 2016, the first CRISPR system that could target and edit RNA was found by Omar Abudayyeh, Jonathan Gootenberg, and Silvana Konermann in Zhang's group, called Cas13a. They then showed in 2017 that this method could be used with isothermal (same temperature) amplification to establish a CRISPR-based diagnostic, which they called CRISPR-Dx, providing rapid DNA or RNA detection with extremely high sensitivity and single-base mismatch specificity. They call their detection platform Specific High-Sensitivity Enzymatic Reporter UnLOCKing (SHERLOCK) and have already used it to detect specific strains of Zika and dengue virus, distinguish pathogenic bacteria, and identify mutations in tumor DNA. In 2020, it received approval from the FDA for detecting the SARS-CoV-2 virus in COVID-19 patients. Perhaps most excitingly, the SHERLOCK reaction reagents can be freeze-dried and used on a piece of paper for field applications, such as during an outbreak.

There is also an ongoing quest to find more CRISPR arrays, which can reveal new biology and also serve as a window into the phage landscape that bacteria are encountering. CRISPR arrays vary in size, but most have an AT-rich leader sequence, then short repeats separated by unique spacers. The CRISPR repeats are usually 23–55 bases, and sometimes they show sequence symmetry, which enables self-folding structures and stem (or "hairpin") loops in the RNA. The size of spacers in different CRISPR arrays is usually 21–72 bases, but there are usually fewer than fifty units of the repeat-spacer sequence in any CRISPR array.

Abudayyeh and Gootenberg's pioneering work, which is similar to Mojica's work in the late 1990s, displays a passion to learn the biology from the sequence data and get as much metagenomic data as possible to find new CRISPR elements and CRISPR arrays. From the Meta-SUB data alone, Abudayyeh and Gootenberg have found >800,000 new CRISPR arrays, matching those of a wide range of species that ride along with passengers on the NYC subways and on others in cities around the world. Some of these new CRISPR arrays and putative enzymes are also being examined by Arbor Biotechnologies and other companies, and the quest will continue to find and apply this new biology as more and more species are sequenced.

DISEASE AND EPIGENOME EDITING

We have only begun to unpack the genetic toolbox that evolution has kindly gifted us. Improvements need to be made to these tools not only to enhance their safety, but also to ensure their functionality within the complex cellular landscape of mammalian cells. Perhaps the most important assumption made in Phases 3 and 4 of the 500-year plan is that the genetic engineering of the early 2000s will *not* be the last to be found, developed, and deployed. In the future, the new and improved genome technologies will take us from an era of genome editing and insertion into one of pure genome writing. Currently, it is estimated that there 1 trillion species on Earth, of which we have only found a few hundred thousand. Further, the amount of genetic data (e.g., bases sequenced, figure 4.2) is increasing exponentially. Thus, more CRISPR

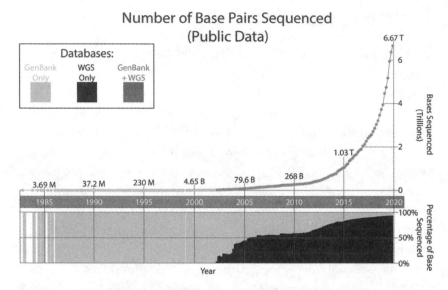

4.2 The growth of DNA data: Number of bases sequenced has continued to grow every year since 1982, visualizing both GenBank (gray) and WGS (black) bases sequenced as reported in www.ncbi.nlm.nih.gov/genbank/statistics/.

methods, or other mechanisms of bacterial or fungal editing, immunity, and genome modification, are likely still to be discovered at an exponential pace from 2020 to 2040. However, even with the currently imperfect genetic-editing toolbox, we can already cure some diseases.

Even now, amid off-target worries and before the pending "error-free era" of perfect genome editing and writing, the excitement of genetic engineering is evident in the explosion of clinical trials using these tools. From 2018 onward, there was a record number of clinical trials started using ZFNs, TALENs, and CRISPR methods (figure 4.3). In 2019, for the first time, a success was secured for two patients with sickle cell anemia and beta-thalassemia blood disorders. Before the treatment, both patients required continual red blood cell infusions to compensate for their improper hemoglobin, which helps deliver oxygen throughout the body. CRISPR Therapeutics worked with Vertex Pharmaceuticals to create an autologous therapy for the disease (cells derived from the same patient, "auto," whereas "allogeneic" means cells originating from a donor). They removed blood stem cells from the patients,

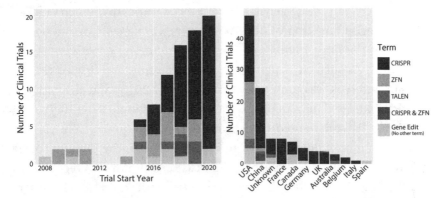

4.3 The types of genetic-editing clinical trials in clinicaltrials.gov: Clinical trials that used or discussed genetic engineering started in 2008, including CRISPR, zinc-finger nucleases (ZFN), transcription activator-like effector nucleases (TALEN), a combination or tools, or nonspecific "gene editing." Left: Editing trials by year. Right: Editing trials by country.

modified them with CRISPR to knock out a gene that prohibits the production of fetal hemoglobin, thus reviving the gene, and then infused the cells back into the patient. After receiving their autologous stem-cell-engineered transplant, both patients had even higher levels of fetal hemoglobin in their blood than clinicians had hoped (>30 percent of cells where only 10 percent may have been good enough). Now, post-therapy, they may live as the first "cured" CRISPR patients. However, their treatment was not easy, requiring a large amount of chemotherapy to kill their own stem cells so the engineered cells could repopulate their immune system. Nonetheless, it is a clear path forward, and sickle cell anemia is not the only disease that can benefit.

At the time of writing, there are over fifty clinical trials using these gene-editing techniques (ZNF, TALEN, CRISPR) for a variety of diseases including cancers, inherited diseases, and blood disorders. Based on this analysis, the United States has the most genome-editing trials employing these tools, but China is closely behind.

Chinese scientists have invested quickly and massively in the idea that we can edit our way out of disease. Within just four years (2012–2016), China went from having no Chimeric Antigen Receptor (CAR) clinical trials (a type of genetically engineered therapy which enables highly precise targeting of otherwise untargetable cells in the body), to having the most number of CAR trials than any other country in the world. As such, it is perhaps no surprise that the first editing of human embryos was announced in China in 2015, and the birth of the first CRISPRed human babies was announced in 2018 by He Jiankui in China. Jiankui wanted to help a young couple who were carriers of HIV to have children that would be immune to the disease. So, he engineered a deletion in a gene called CCR5, a prominent receptor on T cells. CCR5 is essentially a cellular door for most strains of HIV; where HIV can enter the cell, but without this receptor, most strains of the virus cannot gain entry.

The CCR5 deletion was first noticed by doctors in Berlin who were treating an HIV patient a bone-marrow transplant. Unexpectedly, his HIV viral loads had dropped to undetectable levels after many months. The doctors then stopped the antiretroviral medications that were treating his HIV, and, even more remarkably, the viral loads *stayed* at undetectable levels for months, then years. This patient, often called the Berlin patient, was lucky because this incredible result was due to an unexpected mutation in the bone-marrow cells derived from his donor—the CCR5 deletion.

But like all things in biology, the benefit of a CCR5 deletion is cell and time and situation dependent. CCR5 can do more than one thing inside a cell, making it pleiotropic (like most human genes), and its role can change depending on cell type, time, and even the *type* of infection. Work from Robyn Klein and others have shown that, while having a fully functioning CCR5 receptor makes you *more likely* to get HIV, it conversely makes you *less likely* to get West Nile virus (WNV), since CCR5 limits the ability of WNV to infect the T cells and prevents infection in neurons and other cells in the body. In other words, a working copy of CCR5 saves lives in one context (West Nile virus), but it can lead to risk of death in others (HIV). Identifying the pleiotropy of all human genes and the situational result of their editing beyond the specific context for which it is engineered (be it to prevent HIV infection or to improve the ability to live in space) will be an active field of research within 2020–2040.

Some of these lessons can be found in the genomes of people today. In a large-scale effort called the Resilience Project, there is an ongoing quest to find people who should be dead, but are not. These genetic "superheroes" have mutations that should lead to a significant disease but they somehow have avoided it. They were tracked and accrued by Rong Chen, Jason Bobe, Stephen Friend, and Eric Schadt at Ichan School of Medicine at Mount Sinai in New York City. Ideally, if mutations are found that can protect people from disease, rather than just correct a mutation that made a disease, these, too, could be gene-editing candidates. In theory, it could be done on the embryos before it is too late.

Already, Bobe, Friend, and Schadt have found new classes of "superheroes" for HIV resistance. There are "elite controllers": patients who maintain HIV at very low levels even though they are infected. There are "long-term nonprogressors," who are clearly infected but never lose their immune system and seem fine. And there are "elite neutralizers," who produce highly potent, neutralizing antibodies to HIV, which are not normally seen. While it is still early days in the genome-sequencing and genome-engineering era, all these data will be useful in the long term to enable not just fixing what went wrong but improving upon what already exists. Every genome from any patient, can be a lesson and knowledge substrate for a future patient, on Earth or beyond it.

The question of *whether* we can edit human embryos and have genome-designed children is moot; we already can. It is now a question of how we do it, or when, or if we never do it again. Given the announcement by the Japanese government in 2018 to allow genetic editing on human embryos, and the clear interest in not only treating but curing diseases, this question has become mainly one of how.

CONTROLLING DNA METHYLATION

The marvelous ability to take the same genetic code and have it selectively activated and repressed is what enables the extraordinary plasticity of cell types during development, as well as the dynamic changes necessary to respond to our environment. The epigenome is the electrical control box of the cell's biology, which also means it can be spooky when things go awry—like a haunted house on Halloween. As with most complex systems, errors can arise and epigenetic changes can also lead to diseases. As evidence of the epigenome's power, some diseases are even primarily classified by their epigenetic states, rather than genetic changes, which might show higher levels of DNA methylation (CH_3, as described earlier), or changes in how DNA is packaged (open vs. closed), or even just changes in the proteins within the packaging chromatin. Leukemia, glioblastoma, and colon cancer can exhibit "hyper-methylation" phenotypes, which can lead to, and even drive, aggressive cancers. Notably, this can occur even in the complete absence of any observable and known causative genetic changes.

Just like the genome-editing methods described above, the epigenome can also be a target for editing, tweaking, and design. Our knowledge of the candidate sites comes from the areas known to change during normal development, stress, or disease. When we start as an embryo, almost all the epigenetic marks are "reset," meaning that almost all the methylation levels are set to zero, thus opening them all up to full potential. Then, one by one, sites are methylated and genes are silenced to create a specific cell type of interest. If you wanted to change one cell, such as a neuron, into another cell, such as cardiac tissue, in principle, you just need to know all the sites to change. This can be done in two main ways: modifying the types of DNA bases (e.g., cytosine vs. 5-methylcytosine) or adjusting the state and types of chromatin (mono-methylation of lysine vs. tri-methylation, open vs. closed chromatin).

A great example of this power is from Rudolf Jaenisch, who showed in 2018 that epigenome editing can be applied to fragile X syndrome and may even be able to cure the disease (at least in mice). Fragile X syndrome is the most common genetic form of intellectual disability in males, which is caused by hypermethylation of the FMR1 gene, which keeps the gene silent (turning off the gene's activity as if you're turning off your radio by flipping its switch). The hypermethylation site is within a series of "CGG" repeats (specifically the 5' untranslated region, or 5'UTR). To fix this disease, Jaenisch's team needed a way to flip its switch—reversing the methylation and reactivating the gene's expression.

Flipping the epigenetic switch required the invention of a new epigenetic editing toolkit. But, as we discussed with DNA editing and CRISPR, we can learn from biology and don't need to reinvent the wheel. Normally, cytosine methylation is controlled by an enzyme called DNA methyltransferase (DNMT), which converts cytosine (C) to methylcytosine (mC). mC, in turn, can get converted into hydroxymethylcytosine (hmC) by an enzyme called TET1, which was first discovered when looking at a genetic mutation called a ten-eleven translocation (TET). Changing mC into hmC enables the "paperwork" required by the cell's "internal auditing process" eventually to convert the base back into the original C, removing the silencing muzzle that DNMT originally added. This is part of the normal cycling of modified bases in the epigenome, which controls when, and how, genes and their regulatory regions are used. Mutations in TET1 and the other TET genes (TET2, TET3) are linked to leukemia and other cardiovascular diseases—underscoring their importance in basic epigenetic regulation.

To edit the epigenome, the TET1 enzyme can be combined with a modified CRISPR system to target specific sites and control methylation. In this case, Jaenisch and his colleagues merged a deactivated Cas9 (dCas9) protein with the TET1 enzyme, and then used a single-guide RNA to alter the FMR1 gene into an active state, restoring the expression of FMR1 in induced pluripotent stem cells (iPSCs). They then created neurons from the modified iPSCs, which showed effectively normal electrophysiological patterns ("wild-type phenotype") instead of the aberrant patterns from the unedited neurons with fragile X. These edited neurons were then engrafted into a mouse brain to test if they will retain the changes they worked hard for once back into their original microenvironment—much like someone training on Mars with increased gravity with hopes of being able to run a marathon on Earth.

The cells were able to maintain their self control—the editing worked! The expression of FMR1 was maintained in the edited neurons inside the transplanted brain of the recipient mouse. But as hard as this was, this specific process would not translate well to patients who currently have fragile X—given it would require the removal, editing, and replacement of all neurons within the patients. Patients probably wouldn't enjoy this idea, considering they're currently using their brains and are, hopefully, fond of their memories. Nonetheless, Jaenisch and his colleagues went on to show that direct demethylation of the CGG repeats in established brain neurons (postmitotic) is possible and can reactive FMR1 expression. Excitingly, this means that regardless of the reason for a patient's disease, be it genetic or epigenetic, it can be fixed.

THE EPI-EPIOME

Just as the epigenome sits "on top" of the genome to control DNA function, the epitranscriptome sits on top of the transciptome to control RNA, and the development of genetic engineering tools has created a new layer on top of all of the omics' control swtiches; essentially the epi-epiome. Editing the epigenome is not restricted to cytosine methylation; we can even edit the scaffolding that holds the DNA—the chromatin. Chromatin is a hybrid structure of DNA and protein that manages the complex problem of packaging 3 billion bases of DNA into a small bundle in the cell that is only a few micrometers in size. This is no small feat. The length of DNA from one cell would measure two meters in length if you stretched it out in front of you. This long DNA not only (obviously) fits inside each cell, but further, it somehow simultaneously has enough room for molecules to access and read it when necessary. This is the equivalent of taking a string that is the length of the world's tallest building (the Burj Khalifa, at 828 meters) and compacting it so that it could fit in the palm of your hand.

This extraordinary packing is important not only for DNA's protection but also for its regulation. The regulation is mediated by proteins that create chromatin, which can be modified and adjusted just like DNA modifications. Like other proteins, histones are encoded in the genome as DNA, then transcribed into RNA, which is then translated into a unique protein by a ribosome. These proteins are called histones, which come together as two sets of dimers (H2A-H2B) and one tetramer (H3-H4), which then merge, to create an octet of proteins in the nucleus of cells (H2A-H2A-H3-H4). Some other histones serve as linkers

(H1 or H5), but most of the action in the nucleus, and thus the focus of most of the epigenetic editing efforts, is on the core "histone code" of gene regulation in H2, H3, and H4. This is the epigenetic version of taking three ice cream cones, each with sets of scoops (4,2,2) of paired flavors, and merging them into a new eight-scoop tower of functional dessert.

Cells have developed many ways to control gene expression. Regulation of how tightly DNA is wrapped around histones alters how accessible genes are and, therefore, also affects gene regulation. Modification on top of histones enables this precise opening and closing of the chromatin. Just like the catalog of DNA modifications, there is a wide array of post-translational modifications (PTMs) that can, and do, occur to histones and other proteins. In particular, the H3 and H4 histones have long winding tails which serve as a fertile ground to tweak and modify with PTMs. These modifications represent a part of the "histone code" and "epigenetic code" for how cells control the activity and role of various genes and proteins. The list of histone modifications is long and spans some familiar concepts such as methylation and acetylation, but also includes changes such as phosphorylation, citrullination, SUMOylation, ADP-ribosylation, and ubiquitination. As is the case for genes, species, and many other modalities of biology, new modification types are still being discovered. Someday, new PTMs or histones may even arise on new planets that we could use back on Earth.

The histone code is essentially an enormous switch board that controls the genome's manifestations into cell types and cellular responses. Much like the large array of colorful switches in a cockpit that control a plane's altitude, speed, and direction when in the hands of an expert—but would mean certain catastrophic failure if left in the hands of a child—histones and epigenetic states of cells require careful toggling. Enzymes with the duty of adding or removing modifications (essentially the pilot flipping switches on and off) are appropriately named, such as histone methyltransferases, for transferring methyl groups onto histones, or histone acetyltransferases, for transferring acetyl groups onto histones. Further, histone modifications have their own nomenclature, which is actually quite straightforward: (1) the name of the histone, such as "H3" for histone 3; (2) where the modification is on the histone tail, such as "K4" for the 4th lysine (amino acid letter K); (3) the kind of modification, such as "Me" for methylation or "Ac" for acetylation; and (4) the number of modifications which are added, such as 1, 2, or 3 for mono-, di-, or tri-methylation. Through modifying specific sites along the tail of various histones, you can (in theory) get a cell to shift from one state or type to another.

But getting these enzymes to do our own bidding was still science fiction until 2015. Then, some of the first work in histone modifications was published from Timothy Reddy and Charles Gersbach, featuring a completely new kind of construct that paved the way for the previous work discussed from Jaenisch's lab. They built a CRISPR-Cas9-based acetyltransferase that again used a deactivated Cas9 (dCas9) protein, but in this case was fused to the active portion of the human acetyltransferase (p300). This fusion protein enabled acetylation of histone H3 at lysine 27 (H3K4Ac, which, thanks to the paragraph above, everyone now knows how to read), leading to significant activation of target genes from promoters, as well as genes far from these target sites (proximal and distal enhancers). Although other dCas9-based activators had been tried before, their acetyltransferase could change the expression of genes from enhancer regions, using only a single guide RNA to target regions of interest.

Reddy and Gersbach showed that their system is actually modular and could be used to edit essentially any DNA or histone modification as soon as they are discovered through fusing their p300 domain to other DNA-binding proteins. Through tweaking various combinations of modifications, you can essentially get a gene to do whatever you want, whenever you want, wherever you want. With these tools, we have essentially become a conscious layer of regulation, sitting above the genome, transcriptome, proteome, epigenome, and epitranscriptome: the epi-epiome.

EDITING EXTINCTION

Inspiration for genetic engineering can come from unlikely places, even including species that went extinct many years ago. While the potential for current cellular engineering tools is vast, the tools of the future will undoubtedly be even more powerful. As is true with the majority of culture, inspiration doesn't just come from the living. It is entirely possible that the answers to these questions will be unlocked from relics of the extinct.

Each summer in Siberia, Russia, an army of hunters and scavengers wander the barren landscape to look for what they call "white gold," which comes in the form of old woolly-mammoth tusks. These tusks and their fragments are sadly sold in markets around the world, sometimes as high as $1,000/kg. Although woolly mammoths (*Mammuthus primigenius*) used to be common in Siberia and even survived the last ice age on Earth, they could not survive human hunters and the ever-changing landscape, with the last mammoths dying around 4,000 BCE.

In 2013, a research team uncovered the almost perfectly preserved carcass of a female mammoth buried in the Siberian permafrost in Yakutsk, Russia. The majority of her body was still intact, including three full legs, her trunk, and part of her head. While the researchers began to move her from where she took her last rest, some thousands of years ago, they noticed a coagulated, chunky, and dark red substance begin to ooze—blood from an extinct animal. For a geneticist, this is "red gold." Carbon dating estimates that Buttercup (that's right, she has a name) lived about 40,000 years ago.

With Buttercup's well-preserved DNA, scientists and researchers, including Dr. George Church, finally had a chance to pursue a long-sought dream: what if we could bring the woolly back? What if we could splice in enough of Buttercup's 40,000-year-old mammoth DNA into a current enucleated embryo (an embryo with the nucleus removed) of an Asian elephant (the closest-living relative) to resurrect an extinct species? And if it works for the woolly, should we then do this for other species?

The answer is yes, at least according to the Revive & Restore Project (RRP), which features a dedicated page to the progress of, and plans for, resurrecting species including the woolly mammoth. There are at least sixty-three known species that have gone extinct since the 1700s, and likely millions more microscopic species that have also gone extinct during the same disruptive period of colonization and climate change. The goals for the RRP are to "preserve biodiversity and genetic diversity, to restore diminished ecosystems, and to undo harm that humans have caused in the past." The lessons from this work could help us avoid future extinctions, including human extinction, and change how we think about planning trips to Mars and elsewhere.

These efforts are also being spearheaded by the BGI-MGI biotechnology company in Shenzhen, China. Hosted there, amid rolling green hills and some casual flamingos, is the Chinese National GeneBank. Adjacent to the entrance, a giant statue of a woolly mammoth greets visitors with a highlighted phrase (in both English and Chinese) on the side of the animal as a nod to the company's long-term plan: "Preserve for our future." Once inside, you can find a detailed summary of the history of biotechnology and genetics, which shows the clear trend toward the genesis of more and more sequence data and the power that comes with that knowledge.

Walking farther, behind the NGS machines that have heralded a new era in genetics, there is an array of sculptures and photos of various animals accompanied by museum-style placards. Under each animal, details are given about their extinction. Effectively, this is a "de-extinction road map," which aims to bring each of the creatures (such as passenger pigeons) back from the dead, but as healthy as, or healthier than, before—not as zombies. While these goals are difficult, and potentially peculiar to some, the road map posits that we can get back most of the creatures that we have lost. At the time of writing, there are none, but it is possible that by 2040 we may have not just one but several species resurrected from extinction and walking around Earth.

Although bringing back an extinct species may seem like the right or even moral thing (by deontogenic ethics) to do, it should not be done lightly. Just as deleting a species from an ecosystem can have unexpected consequences, so can reintroducing a species that has been extinct for thousands or millions of years. It is possible that having woolly mammoths wander the Siberian plains will be good for global warming, because these creatures can graze on the grasses and decrease the speed with which more carbon is emitted as the permafrost thaws. But woolly mammoths might also disrupt the ecosystem in other ways, such as by serving as a reservoir for viruses that were long extinct, trampling over the small homes of mammals that live in the plains, or releasing methane from their own ancestral flatulence. As with most of biology, tests will be needed in the short term to see if we think any ecological disruption is doing more harm than good so that we can adjust accordingly.

While controlling the evolution of the past, present, and future seems scary and wrought with incredible hubris, the reality is that we *already* have been engineering and modifying species and the environment around us, except previously we were doing so by accident with no foresight. Now, finally, it can be done with a sense of responsibility and purpose. This self-directing evolutionary process is a necessary agency in the development of humans that will enable us to fulfill our purpose. Thus, rather than running from this duty, out of concern that we might "screw it up," we actually have an ethical obligation to run *toward* this duty and take ownership as the only species with the capability to preserve all others. Given enough time, doing nothing is the only sure way to guarantee extinction of all life. We must act.

If you asked Buttercup her opinion, she would likely agree.

EMBRYO EDITING AND DESIGN

The era of 2021–2040 will be dominated by the discovery of new cellular engineering tools, improvement of current technologies, and the utilization of these systems to cure complex diseases within in vitro and in vivo settings in preclinical and clinical models. As is the case with any revolutionary, but potentially risky, technology, there may also be a handful of "genome resurrections" and "reboots" where tools and functions from one species are moved to another. Finally, these tools will, inevitably, also be pointed to the very first cell of an embryo.

Yet adapting cellular engineering technologies to edit embryos will likely be a slow process, as it should be. At the International Summit on Human Gene Editing in December 2015, the organizing committee issued a statement about the appropriate uses of these technologies. Specifically, they stated that it "would be irresponsible to proceed with any clinical use" that would entail "making genetically modified children, unless and until (i) the relevant safety and efficacy issues have been resolved and (ii) there is broad societal consensus about the appropriateness of the proposed application."

In 2019, the same year Dr. He Jiankui was sentenced to prison for creating CCR5 CRISPR-edited children, many of the scientists who pioneered CRISPR methods (Lander, Baylis, Zhang, Charpentier, Berg, and others) called for "a global moratorium on all clinical uses of human germline editing—of changing heritable DNA (in sperm, eggs, or embryos) to make genetically modified children." They correctly noted the issues of balancing selection, pleiotropy, and the incomplete state of our knowledge of genetics and cellular biology. Playing God with a system of biology of which we do not know all the components, or for which we may not have a complete model, is risky.

Even the seemingly benign idea of reducing disease risk by modifying genes with other naturally occurring alleles can have hard-to-predict impacts. For example, variants in a gene called SLC39A8 can decrease a person's risk of developing hypertension and Parkinson's disease, but also increase the risk of developing schizophrenia, Crohn's disease, and obesity. Just as there is the trade-off of West Nile protection or HIV infection from CCR5, there is a trade-off with the gain or loss of many other genes and mutations. There is no free lunch. Maybe an organism can get a free cookie now and again in a transposable element, but even that is rare, and the trade-offs and our staggering ignorance of biology are evident and exhibited at myriad scientific and clinical conferences.

Nonetheless, this has not stopped the National Academy of Sciences and the National Academy of Medicine from seeing the possible benefits and even endorsing the idea of germline editing. In 2017, they stated that "with stringent oversight, heritable germline editing clinical trials could be permitted for serious conditions, but that non-heritable clinical trials should be limited to treating or preventing disease or disabilities." They recommended a clear set of guidelines for moving forward, including: (1) absence of reasonable alternatives; (2) restriction to editing genes that have been shown to cause or predispose to a serious disease or condition; (3) credible preclinical and/or clinical data on risks and potential health benefits; (4) ongoing, rigorous oversight during clinical trials; (5) comprehensive plans for long-term multigenerational follow-up; (6) continued reassessment of both health and societal benefits and risks, with wide-ranging, ongoing input from the public; and (7) reliable oversight mechanisms to prevent extension to uses other than preventing a serious disease or condition.

These guidelines have also led to a proposed hierarchy of testing, which are basically the steps for ensuring safety in genetic engineering. Some of these discussions are ongoing at international scientific meetings and many advisory bodies, such as the US National Academy of Medicine, the US National Academy of Sciences and the UK Royal Society, led by Drs. Kay Davies and Richard Lifton, as well as the Genome Project-write (GP-write) meetings, led by George Church, Jef Boeke, and Andrew Hessel. The Scientific Executive Committee (including myself) and members of the GP-write consortium are constantly debating and preparing for this future. The proposed hierarchy for steps in genetic engineering includes an iterative process that could be (1) animal models, (2) in vitro human cells, (3) dogs, (4) primates, (5) human cells, and, finally, (6) humans. But even with this paradigm, because of pleiotropy and varying developmental trajectories of cells (some of which are known, most of which are not), these therapies must be tested step by step, compared across many different genetic backgrounds (ancestries), and also across development. This includes maps from the first cell of embryogenesis onward, in a test tube or a womb, inside of a body or out.

"TEST-TUBE" BABIES

Most of the work on editing and modifying human embryos has been gathered by clinicians and researchers using in vitro fertilization (IVF) methods. The first IVF baby (Louise Brown) was conceived in 1977 and born in July 1978, a product of cutting-edge science and the need to solve a medical problem for her mother (Lesley Brown), who had fallopian-tube obstruction. Louise had a child of her own in 1999, and thus proved that IVF babies could function and reproduce just as "normal" children can. Showing that assisted reproductive technologies (ARTs) can create healthy people who can make their own babies helped quell fears that somehow the IVF babies would be "deficient."

This is especially relevant because in 2021, about 2 percent of all babies born in the United States were born with IVF.

However, the extent of ART safety and associated disease risks are still being studied. For example, during the IVF process, strong medications are used to induce ovulation, embryos are cultured outside of the body, frozen and thawed, and large doses of progesterone are normally used to help with development—all processes that an embryo in a womb wouldn't typically face. Also, intracytoplasmic sperm injection (ICSI) directly injects sperm into the ooplasm, which removes the selection process that normally occurs on the oocyte membrane (poor swimmers) to decrease the chance an abnormal, and potentially sick, sperm can be the genesis of new life. This has led to a few studies linking IVF and ICSI children to a risk of higher blood pressure or insulin resistance, but so far there have been no clear red flags indicating that we should abandon the process.

Defective mitochondria (the "power house" of the cell) can be devastating to an embryo, resulting in death before birth or seizures, pain, and a shortened life expectancy if born. About 1 in every 4,300 individuals in the United States has a mitochondrial disease (e.g., Leigh's disease). Children inherit almost all of their mitochondria from their mother's egg, which enables two options to fix these issues: either repair the egg before fertilizing or repair the embryo itself (after fertilizing). In "embryo repair," the donor eggs and mother-to-be's eggs are both fertilized with the father-to-be's sperm, making two viable embryos. The pronuclei (the cell's transient nuclei) are removed from both embryos, with the donor's pronuclei destroyed and the mother-to-be's pronuclei placed into the donor embryo. Then, the embryo can develop as a normal embryo. In "egg repair," an egg donor with healthy mitochondria and the mother with unhealthy mitochondria have their eggs placed in a dish. First, the nucleus from the mother-to-be is isolated, containing her human genetic payload. Then, the donor nucleus is destroyed, and the mother's nucleus is inserted into the donor egg, which still has the complement of healthy mitochondria. The egg, which now contains the mother's nucleus, is then fertilized by the father's sperm.

Both of these strategies can be used to create healthy, "three-person" babies. In 2015, the United Kingdom passed legislation to enable the birth of these "three-person" babies. The first medical license was granted in 2017, and the first two subjects were selected in 2018 to treat myoclonic epilepsy with ragged red fibers (MERRF syndrome). Before that procedure, the normal development of the child could only be hypothesized, though now it is an established and reimbursable medical practice.

ARTIFICIAL WOMBS

The next big technological advancement of embryo design, editing, and selection would be the creation of an "artificial womb," where an embryo can grow and develop to full term. Also known as "exowombs," artificial wombs were first developed to enable premature babies to survive after emerging too soon from their mothers. The first example came in 1996, when Yoshinori Kuwabara and his research team wanted to reduce the risks of death that premature babies face. They used fourteen goat fetuses removed by cesarean section after four months of normal gestation, then placed them in their exowomb chamber, linked to an umbilical cord and their synthesized placenta. The team emulated the fluids, nutrients, and temperature of a pregnant goat in the artificial amniotic fluid. While most of the goats died, a few survived up to three weeks, reaching effectively full term, but they all had deformities or lung problems.

In 2003, a mouse embryo grew almost to full term in an exowomb made by Helen Hung-Ching Liu, the director of the Reproductive Endocrine Laboratory at Cornell University's Center for Reproductive Medicine and Infertility in New York City. She had first used sheets of human tissue composed of cells from the endometrium, which is the lining of the uterus, but they were not thick enough. So, she built better, three-dimensional networks of tissues resulting in an exowomb that better resembled the actual womb. Mouse embryos attached, formed blood vessels, and grew—mammal exowombs were no longer science fiction.

In 2017, scientists at the Children's Hospital of Philadelphia expanded these ideas to go beyond the mouse and to even bigger animals—lambs. They developed an exowomb for fetal lambs, which was essentially a plastic bag filled with artificial amniotic fluid. The lambs' umbilical cord was attached to a machine that served as a placenta and provided oxygen and nutrients, while removing biological waste. The researchers kept the machine "in a dark, warm room where researchers can play the sounds of the mother's heart for the lamb fetus." Like the Japanese goat research in the 1990s, the system succeeded in helping the premature lamb fetuses "develop normally for a month." While early in its implementation, the procedure is a proof of concept to aid the development of premature babies and is a first step in the ability to bring a child to full term starting just at the single cell, embryo stage, in vitro.

The closest we have gotten to a portable womb was not synthetic, but instead derived from a donor. In October 2014, a Swedish woman who had received a "womb transplant" gave birth to an IVF baby. The mother-to-be was originally born without a uterus and asked her friend, who was in her sixties, if she could have her uterus since she was no longer using it (she was already postmenopausal for seven years). The British medical journal the *Lancet* reported the success, showing that a woman born without a uterus (as is the case for women with Mayer-Rokitansky-Küster-Hauser syndrome), or a cancer patient who has had her uterus removed (e.g., hysterectomy), could imagine getting a uterus transplant and giving birth to her own baby. Good friends share clothes; great friends share organs.

Another obvious application of exowombs is to help couples who cannot conceive on their own. This procedure would have to be as safe as, or safer than, that of using surrogate mothers or donor wombs. While some women are anxious about having other women carry their babies, surrogacy is (for now) the safest way. But other complications emerge for the surrogate mothers, who risk their life and livelihood as biological incubators for other people, and whose compensation can range from a few thousand dollars to fifty thousand or more. There is already a market for surrogacy, and it would be disrupted by artificial wombs.

If we reach an age where artificial wombs are commonplace, other complications for society emerge. The first issue is most critical for the United States, where the law securing a woman's right to an abortion stems from the landmark 1973 Supreme Court case *Roe v. Wade*. This law was framed primarily upon the fact that a fetus could not survive outside the mother's body before twenty-eight weeks, as well as for the protection of the mother's body and ability to choose. By 2021, it had already become increasingly common to see twenty-four-week-old babies in the incubators at hospitals, further reducing the time needed for a child's viability outside of the mother's body. As this incubation period continues to decrease with exowombs, it could eventually reach zero, and then every embryo could be brought to full term. Such a technological shift could potentially undermine the *Roe v. Wade* case and shift the control of the embryo from the mother to the state, similar to the dystopian novel *The Handmaid's Tale*. Or, positively viewed, it could enable every woman to choose exactly when, how, and how long pregnancy would progress and ensure that almost no fetus is lost.

Another issue concerns the developmental aspects of fetal and child development. Work from Janet DiPietro at Johns Hopkins University has shown that pregnancy is a two-way street of constant communication between the child and the pregnant mother, which would be ostensibly lost in the exowomb. DiPietro has shown that a child can respond to its mother's moods, that fetuses react almost instantly to changes in maternal position or emotions, and that the fetuses continually communicate as well, teaching their mothers to pay attention to them while shuttling out hormones and chemicals. There is a biochemical highway connecting mother and child at all times, which may be hard to replicate and would likely be lost, at least in the first version of the exowomb.

Thus, some sense of nostalgia may occur if we "lost" this connection and if everyone switched to using exowombs. Yet, current birth and maternal health data show that the current biological process of human gestation and birth is not ideal. Even with modern medicine, hundreds of women die in childbirth each year in the United States, and thousands around the world. The human head cannot get much bigger and still emerge through the vaginal canal, a fact that is helping to propel the rise of more C-section births and procedures. Also, the incidence of preeclampsia and prenatal complications is rising, and pregnancy can result in gestational diabetes that does not always resolve at birth. Clearly, pregnancy and birth are not always an easy biological road to travel. It can be improved; a new liberty can emerge.

VITALISM AND NEOVITALISM

Some people may still resist the idea of an exowomb and not carrying their child in their own womb because they view birth and pregnancy as possessing an inherent magic. Yet, this is not be the first time this type of argument has been made. In the 1500s, a movement in science called *vitalism* posited that there was something so unique about life that it could never be reduced to a molecular deconstruction. They believed that living organisms are so fundamentally different from nonliving entities that they are governed by different principles and must possess some kind of nonphysical element—a kind of "soul" given at conception and embedded into every fiber of one's being. According to that idea, no organic compound could ever come from an inorganic source.

But that belief turned out to be wrong. In 1828, a German chemist (Friedrich Wöhler) used silver isocyanate and ammonium chloride to synthesize urea completely from scratch, instead of waiting for it to come from the human body; an organic compound created from inorganic sources. Since then, cloning, embryonic stem cells, and IVF have done wonders to improve our understanding of development and have not revealed any sign of a vital force that mediates this process—only new findings which help us continually understand these processes. Indeed, work from Ali H. Brivanlou and Eric Siggia at Rockefeller University has shown that human embryos continue to develop just fine in a dish, all alone for fourteen days. Systems biology has shown that even complex dynamic states in many organisms can be modeled and predicted. Both the human womb and synthetic exowomb are guided by chemistry, physics, and biology, not magic; both wombs can create and save lives.

The human womb has been a wonderful result of unguided evolution, which has brought us to where we are today. But, like everything else in biology, it can be improved through purposeful engineering. It can be improved. Any risks associated with pregnancy (including increased risk of stroke and heart attacks) and the high physiological stress that pregnancy puts on the human body could be lessened, or removed, with the help of exowombs. For example, higher risk of breast and other cancers is present for women who give birth at an older age (>30). This is mostly due to the exposure to fluctuating hormones (estrogen and progesterone), which drives cell differentiation, growth, and potentially oncogenesis. In fact, the older a woman is at the age of her first full-term pregnancy, the higher her risk of breast cancer. Simply put, women who give birth for the first time when they are over thirty years old have a higher risk of breast cancer compared to women who never give birth. Also, women who have recently given birth have a short-term increase in breast-cancer risk, but that declines after about ten years, likely also relating to the surge of pregnancy-related hormones.

Conversely, there are some protective aspects to cancer risk that come with pregnancy. For example, work from Bernstein and colleagues showed that women who give birth before age twenty have about half the risk of cancer compared to women whose first become pregnant and give birth after thirty. Also, women, especially younger women, who have given birth to five or more children have half the breast-cancer risk of women who have not given birth. Further, women who have had preeclampsia or breastfeed for an extended period (at least a year) may have a decreased risk of developing breast cancer.

Through a better understanding of the biochemistry of pregnancy, we could engineer the ability to give women and their children the best of both worlds—a joint treatment that provides the protective hormonal and cancer-decreasing benefits of pregnancy, as well as eliminates the oncological risks. In such a paradigm, we could theoretically engineer a device to simulate the biochemical response of pregnancy for a woman, which could be linked to an exowomb in which their baby develops. If the baby kicks in the exowomb, the woman (or man, or other) could feel it as well with an attached device, and they could be intimately, wirelessly connected. The success of such an engineering marvel would depend on the ability to recreate, and improve upon, in vivo human development and allow a human to gestate another human from afar. In this setup, the "remote gestator" would be able to obtain all of the protective benefits of giving birth without actually having to have a child. This plan rests on the hubris of engineering and medical monitoring, but preliminary data in animal models and epidemiology point to a future where it is possible to improve pregnancy and reproduction for both the gestator and child.

LOOKING AHEAD

After several decades of progress, there will be an ever-shrinking number of diseases and disabilities that are either untreatable or that require life-long medical intervention. Eventually, any cell will be able to be used to become any other cell in the human body (by genetic and epigenetic tweaks), synthesis costs for DNA will become affordable for genome writing (and not just editing), and artificial wombs will be safer than live pregnancy for both the mother and child. Taken together, these technological and biological marvels will bring humans into a new era of health, safety, and longevity.

In this ideal world, the accumulation of genetic tools gleaned from all organisms that live, will ever live, or will ever live again (through genetic resurrection) can be employed to decrease risk and improve the livelihood of any other organism. Just as a child can donate toys to other children they want to help, organisms can and should share their unique abilities and evolutionary adaptation strategies with each other in a directed, helpful process. This sharing and learning would continue as we look to longer space missions, including long-term plans for monitoring the success or risk of biological engineering as we send people to those planets.

5

PHASE 3: LONG-TERM TRIALS OF HUMAN AND CELLULAR ENGINEERING (2041–2100)

> [Evolution,] a process which led from the amoeba to man appeared [to some philosophers] to be obviously a progress, though whether the amoeba would agree with this opinion is not known.
>
> —Bertrand Russell

By 2040, editing genomes and epigenomes will be common, safe, accurate, and affordable. Further, we will be able to selectively direct the differentiation of specific cell types in vivo, as needed, for therapeutic and clinical purposes. These improved abilities will enable the creation of "protected genomes." Once this has been established, we will quickly apply these methods to preventative safeguarding for astronauts, including mission-dependent protection strategies on long-duration spaceflights. These technologies will likely become widely adapted and embedded within society itself. Some of this is already happening in the early twenty-first century.

GENE THERAPY TO REVERSE BLINDNESS

Scientists across countless disciplines, since the identification of the genome, have been determined to make genetic editing safe, common, and easy. In the early twenty-first century, the NIH substantially helped scientists in their quest to achieve this goal. The NIH's set of financial and scientific resources, called the Common Fund, devotes large financial resources to projects that can quickly move a field forward, including "high-risk, high-reward" grants and programs for scientists and clinicians. One of the Common Fund's key projects, the Somatic Cell Genome Editing (SCGE) program, was launched in January 2018, with the aim of improving the efficacy and specificity of gene-editing approaches in order to help reduce the burden of common and rare diseases caused by genetic changes. This is a well-financed and well-organized way to "right the wrongs" of genetic diseases. The SCGE funds are used to develop cell-specific and tissue-specific delivery vehicles to create more precise genome-editing tools.

If only one cell type (e.g., cardiac cells) needs to be edited or modified, then specificity is paramount. A genome editor intended for the heart could potentially disrupt neurons in the brain or nephrons in the kidney if inappropriately aimed. As described in the previous chapter, there are key technical limitations with genome-editing technologies—especially in their delivery. Most base-editing enzymes like CRISPR are too large to fit into AAV vectors, meaning we must use other techniques which are currently less commonly used in the clinic to get them into the necessary cells. Another challenge is the precision and control of the editing. If the widespread temporary change in the expression of a gene is required, potentially augmentable through epigenetic editing, then the fleeting expression of the genome-editing machinery may be all that is necessary. But if a gene therapy requires the addition of a specific gene into a specific cell and can potentially inappropriately alter the function of other cells, then the editing technology needs to either only embed into the cell type of interest or, at least, only be expressed within this cell type.

A great example of cell-type specific, in vivo genetic engineering is in the eyes. The first CRISPR treatment for a form of blindness called Leber congenital amaurosis (LCA) was approved by the FDA in December 2018, for Editas Medicine and Allergan. LCA is the most common cause of inherited childhood blindness, appearing in about 3/100,000 births, and historically, there was no treatment or cure. Patients with this disease usually only see extremely bright lights as blurry shapes, and most patients eventually lose all sense of sight. There are various kinds of LCA, which result from various mutations in a subset of genes (e.g., CEP290, CRB1, GUCY2D, and RPE65). Each of these mutations will then likely require a different treatment design to specifically target and treat the patient. This LCA CRISPR treatment targets one of these mutations as a proof-of-concept and holds the potential to offer a full cure.

The new CRISPR therapy built upon earlier work in gene therapy. The first AAV LCA therapy (called Luxturna) was approved by the FDA in 2017 for treating LCA2, in which the virus carried the payload into the retinal cells to replace the defective gene. Luxturna seemed to work fine with no known side effects, and a related trial in the Netherlands enabled better vision for 60 percent of study members. Now, with both a viral vector and CRISPR editing, in vivo modification of genes in specific cells has made it possible to repair an inherited genetic error.

This extraordinary era of somatic and therapeutic genome editing is only the beginning. Once we can validate the efficacy and safety of these procedures, we will be able to directly edit otherwise fatal or debilitating mutations within an embryo before birth. Mutations can be closely monitored during development and corrected when necessary to improve the chance of survival for the baby and also a likely improved quality of *life* for this individual. Indeed, highly stage- and cell-specific treatments represent the best drug-to-patient match possible for therapies.

GENE THERAPY TO REPROGRAM CELLS

But what if a person's ailment isn't genetically driven? Or what if the genetic abnormality is highly complex, but results in the long-term loss of a given cell type? For example, what if one's eyesight is lost during one's lifetime, but we know which cells have been lost? Here, too, highly specific treatments can be designed to treat a given disease for a specific patient. Ideally, scientists could reprogram other cells to pick up the slack and restore cellular functions.

Indeed, such an idea is possible. In the human eye, the retina is home to about 200 million (M) "rod" neurons, which help the eye determine if there is light versus darkness. A smaller subset of cells (~5M) are the cone neurons, which discern colors and patterns in the field of view. Work from Jeffrey Mumm in 2018 showed that in an animal model (zebra fish) for blindness, where the cone neurons were damaged, the color vision could be re-created by modifying the rod neurons and turning them into cone neurons. This was accomplished by multiplex CRISPR (where multiple different areas of the genome are targeted at once) of the retinal cells resulting in converting rod cells into cone cells. In the human eye, this would be an easy trade-off; if you reprogram 5 million rod neurons, there are plenty of rod cells to spare (195 million). Thus, a procedure can use reprogramming, instead of editing, to leverage existing cells as a reasonable new cure for macular degeneration and blindness.

But why not have the best of both worlds? Why not retain all 200M rod cells and still regain cones? With enough cellular reprogramming acumen, it would be possible to redifferentiate cells and allow them to asymmetrically differentiate (i.e., divide into two different cell types). Within this paradigm, one could use the cells which already exist in a tissue, allow for the development of a new cell type of interest, and still retain the original cells that were present.

Once these highly complex therapies have shown both efficacy and safety on Earth, we can begin using them in simulated Martian environments, Mars space stations, and even on Mars itself. These types of technologies are crucial to address issues which may come up when far from Earth, given the large range of different conditions in which they can be used, by simply changing a target sequence or delivery system. Instead of needing an entire pharmacy where each drug has a different target and use case, you can have a singular system with one modular component (which can be synthesized) and then adapted to whatever need arises.

There is a planned space station for orbiting Mars, called Mars Base Camp, that is slated for orbit by Lockheed Martin by 2040. While "Mars Base Camp" might sound a bit like a goofy summer camp with a crimson theme, it is far more than that. It is the place where things begin to become especially interesting for the questions about human persistence on other planets and the means by which we can survive, long term, on other planets. We could use cellular reprogramming to adjust our needs as they arise. To do that, we need to first know what we are up against.

GENETIC DEFENSES FOR SPACE

When we travel to Mars, we will be beyond the Van Allen belts and the protective magnetosphere of Earth. These belts provide Earth and all its creatures extraordinary protection from the onslaught of solar and galactic radiation that would otherwise bombard us daily. Earth's constantly spinning molten iron core creates this protective electromagnetic shield deflecting the oncoming radiation away from us. This field is, quite literally, an invisible shield and is a large part of the reason Earth's current life exists at all.

You will get in trouble if you stand naked on the surface of Mars, although not for legal reasons. This trouble will come in the form of cold exposure of the skin, as well as the radiation exposure due to Mars's current lack of a magnetosphere. Mars did once have a molten core and a protective magnetosphere (as well as abundant and flowing water), where your favorite body parts would have been protected, but the planet's iron core has mostly stopped moving. Thus, astronauts not only have to face the harsh radiation exposure while living on the surface, but also on their journey to the beautiful red planet. These radiation levels will be unlike anything anyone has ever faced on Earth or on any other space mission as of the twenty-first century (figure 5.1).

The US Nuclear Regulatory Commission (NRC) assumes that any dose above zero can increase the risk of radiation-induced cancer, meaning that there is no truly "safe" threshold for exposure. As astronauts get exposed to more and more radiation, the cells of their bodies become progressively more damaged because not all DSBs, rearrangements, or mutations are getting repaired. Also, while severely damaged cells can self-destruct to save the rest of the body, the mechanism of apoptosis is imperfect and fails more and more as we age. All these "molecular insults"—the cellular, epigenetic, and genetic changes and damage—increase the long-term risk of developing cardiovascular disease and other diseases, such as cancer.

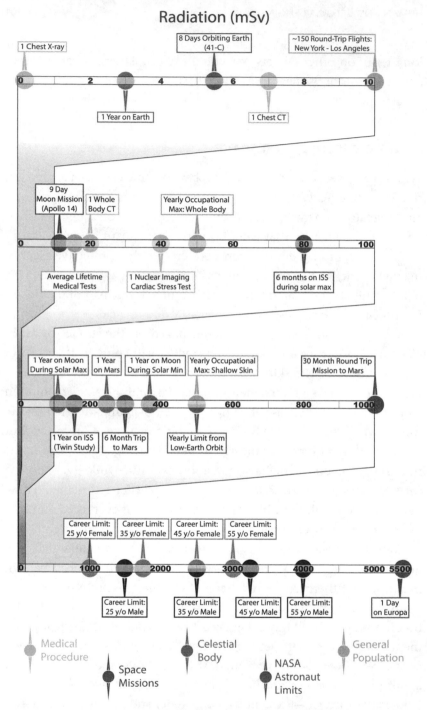

5.1 Radiation risks and levels: Estimated levels of radiation for varying-duration missions and exposures, including medical procedures, the impact while on various celestial bodies, specific space missions, NASA-recommended astronaut limits, and the general recommendations for terrestrial radiation. (See color plate 10.)

To measure the biological impact of radiation on humans, we use the unit Sievert (Sv). Sv is defined such that 1 Sv (or 1,000 mSv) is equivalent to a 5.5 percent increased chance to develop cancer—one of the main risks of radiation. The exposure of 1 milligray (mGy) (the *physical* amount of radiation) results in the *biological* effect of 1 mSv (.001 Sv). The study of radiation is called "dosimetry," and it sometimes uses other units such as the roentgen (R, which equals 10 mGy), which can also be expressed as a rad, or the "radiation-absorbed dose." Radiologists love backup units, apparently.

While Scott Kelly was orbiting Earth 400 km above Earth's surface, moving at 8km/s, he was also receiving the equivalent of four chest X-rays a day (a chest X-ray is approximately 0.1 mSv), and he was exposed to ~0.43mSv a day. This daily radiation is far less than what the Apollo astronauts experienced (13–16 mSv) when traveling to the moon and back, and less than the amount of radiation from some nuclear imaging, such as cardiac stress testing (40 mSv). However, by the time Scott returned to Earth, his dosimeter (the device that measures radiation exposure) showed a total of 146 mSv, representing a relatively high total dose of radiation. But it is also important to note the difference between an instant dose (figure 5.2) and radiation exposure over a longer duration of time. A large dose in a short time is more dangerous than the same dose over a longer time, since the body has time to heal.

Epidemiologic studies have found that one's chance of dying from cancer, as a result from radiation, is also a function of where a person lives. The higher the elevation—such as mountains—have less

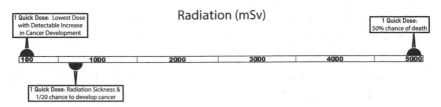

5.2 Radiation risks from a single dose: Expected health consequence from a single dose of radiation at varying levels.

atmosphere above them and thus less protection from radiation coming from space. Yet radiation can also come from below, from the soil and stones around us, which contain uranium, radon, thorium, and other natural sources of radiation. For example, the worldwide average natural dose to humans in the United States is about 2.4 mSv per year, which is four times higher than the worldwide average exposure (0.6 millisieverts). The Rocky Mountains show threefold more radiation than the national average. All of these are small, relative to the radiation of a year in space, as seen with Captain Scott Kelly. According to the NRC calculations, the year-long mission for Scott increased his absolute lifetime risk of cancer by 0.58 percent.

Fortunately, all NASA astronauts have excellent medical care, and most participate in a longitudinal health study to identify and track risks and physiological changes. Our ability to map mutations and understand their potential impact has dramatically improved through using next generation sequencing in temporal studies across a range of methods like those used in the NASA Twins Study. These somatic mutations can affect slowly dividing cells, such as hematopoietic stem cells (HSCs), as well as their more rapidly dividing progeny. Sometimes, mutations in HSCs enable a fitness advantage and a "clonal expansion" of the mutant blood cells; some of the first evidence of these mutated clones in otherwise normal individuals was found by Dr. Ross Levine at Memorial Sloan Kettering Cancer Center (MSKCC) and the Mason lab in 2012.

Extending this work, in 2018, Drs. Duane Hassane, Gail Roboz, Monica Guzman, and colleagues showed that these mutations could forecast the development of cancer and cardiovascular diseases sometimes fifteen years before they occurred. Effectively, the mutated cells were a ticking time bomb; they already had a "first hit" of the two hits needed to develop cancer (known as the "two hit hypothesis"). Indeed, this new field of study around the slow accumulation and risk of clonal mutations in the blood, "clonal hematopoiesis," has led to a better understanding of mechanisms of risk and inflammation in our blood. Such tracking of mutation burdens, identifying their associated risks, and implementing interventional therapies when necessary will be essential to safely explore Mars and other planets. We are all mutants, to varying degrees; it is only a question what kind of mutant we are. And on which planet.

This slow, inexorable march toward molecular oblivion and endless progression of accreting mutations in the blood might seem like a runaway train that cannot be stopped, but it is not. Not only will we eventually be able to monitor this process and intervene when necessary, but there is also increasing evidence that this process can be slowed or perhaps even stopped. Mouse models studied by Iannis Aifantis at New York University and Omar Abdel-Wahab at MSKCC showed that vitamin C can reduce the proportion of the mutant clones which appear in the blood (called the variant allele frequency, or VAF). These data showed that preventative therapies can help lower, or even potentially avoid, the risk of leukemia and cardiovascular diseases.

Interestingly, when we analyzed purified cell fractions from Scott Kelly's time in space, we found that he carried some mutated clones in the TET2 gene. The overall VAF for mutant clones decreased, resulting in him having a younger "blood age" after his year-long mission. Given his stable "epigenetic age" and longer telomeres, as previously discussed, this may not seem surprising. However, after looking at his blood again in 2020, the VAF not only came back, but was at even higher levels than before. Also, Mark Kelly had distinct mutations from his brother, which changed at a different rate, indicating distinct trajectories of mutations in the blood of the twins. When I presented these data to both astronauts in 2020, a question arose: "Can you just CRISPR the mutations away?" I had to answer, "In mice yes, but in humans . . . not yet."

This underscores the importance of deep and continual monitoring of astronauts to build a basis for comparison in future studies, as well as rigorous testing of genome-editing methods. Building genomic defenses for space includes first identifying what can go wrong (e.g. clonal selection of radiation-induced mutations), how to identify it (continual monitoring of VAFs), how to intervene (genetic editing to remove the mutation or cells which are mutated), and when to intervene (if the mutation is associated with increased chance of cancer or other risks). Ideally, these defenses are preventative, as opposed to reactive, through including medications or nutrients such as vitamin C, as discussed above, or highly specific genetic or epigenetic editing in a mission-specific context.

BEYOND EARTH

These metrics of clonal hematopoiesis and even Scott Kelly's response to a year on the ISS are still within the safety of Earth's magnetosphere. As one wanders farther away from Earth, into deeper parts of the solar system, and outside of our solar system itself, risks become progressively worse. Spending a year on Mars would expose an astronaut to around 250 mSv, and a thirty-month mission to and from Mars would likely expose a crew to about 1,200 mSv. The actual amount of radiation is based on the amount of galactic cosmic rays coming from long-dead, distant stars, as well as solar flares from our own star which is largely dependent on the sun's natural eleven-year cycle between solar maximum and solar minimum—visible by the activity on the surface of the sun through sun spots.

The expected level of radiation exposure from a *single* Martian mission starts to approach the *career* limits for astronauts, which are currently based on both the age and sex of a given astronaut. Of note, these current limitations are based on the Japanese atomic bomb survivor life span study, which, at the time, was the most comprehensive study of the long-term consequences of high-dose radiation exposure on the human body. From this study, women were identified as having a higher chance of radiation-induced cancers. This was not only identified in female tissue-derived cancers relative to male tissues (e.g., breast and ovarian versus prostate) but also for lung cancer, where women showed a two to three times higher risk. However, this study is far from ideal to truly assess risks of spaceflight on the human body.

As an example, the career limit for twenty-five-year-old women is lower (1,000 mSv) than their male counterparts (1,500) (see figure 5.1). The maximum cumulative exposure increases as people age because their bodies have more time to heal from the insults of space and their remaining life years are, of course, shorter. For example, the career limit is much higher for a fifty-five-year old female (3,000 mSv) and male (4,000 mSv) than younger astronauts, which again references the difference in the result of radiation exposure as it is spread over time. To improve our models, additional studies are currently being conducted—such as the NIH's Million Person Study—which will be able to better understand the long-term health consequence of longer-term exposure to lower doses of radiation. Of course, the best comparison will always be astronauts themselves, on whom we will continue to gain more information in Phase 3 (2041–2100), especially as space begins to open to tourists and more long-duration missions are launched from NASA and other space agencies.

Yet the radiation levels from Hiroshima, Nagasaki, and for astronauts in low Earth orbit pales in comparison to what would happen on some of the harshest missions, like Europa. The massive gravitational forces of Jupiter create a wildly fluctuating magnetosphere that almost rivals that of the sun. Trapped particles near the poles of Jupiter are concentrated, accelerated, and blasted out at the nearby Jovian moons. Standing for one day on the Europa's icy surface would expose an astronaut to 5,500 mSv. To put this in context, hematopoiesis (the blood cell–forming process in a person's bones) essentially halts at 500 mSv, and cataracts in the eyes can occur at 5,000 mSv. From one day of this exposure on Europa, at this rate, the chance of death within thirty days is about 50 percent. Clearly, we need some protection.

People tend to like planets (and moons), for many reasons. One good thing about these celestial bodies is that they serve as a giant shield, providing "free" protection from cosmic radiation. Half of the planet protects you while standing on the surface because all radiation coming from below your feet has to first get through your planetary protector. Further, planets can give a free protective blanket or shield in the form of thick atmosphere or regolith (soil), which you could dig through to get under the surface. However, if external protections are not available, or simply don't offer ideal long-term solutions, then we can build internal, genetic protections within our cells.

GENETIC OPTIONS FOR RADIATION DEFENSE

As described in chapter 4, lessons about (and substrates for) genetic defenses can come from any organism, ranging from the extra copies of p53 in elephants to the Dsup gene in tardigrades. We will continue to learn from extremophiles and their unique adaptive strategies, which will undoubtedly provide inspiration to direct adaptation to never-before-experienced, harsh environments (including new planets). Lessons from previous, current, and future space missions will lead to a better understanding of human response to spaceflight and reinforce our current frameworks of radiation response and resistance.

The first ideal genetic radiation resistance framework would be through the combined engineering of TP53 and Dsup. As previously discussed, this would require highly specific engineering and dosage compensation studies to enable radiation resistance without potential negative consequences of simply having too much TP53 expressed within a cell. Once the ideal regulatory framework has been designed to control for the overall level and timing of their expression, an iterative approach could be employed to continually add in other genes— one at a time—and identify how the cell responds. Radiation resistance could come from many other genes, some of which we have yet to find in nature, whereas others may be found through studying what is active in astronauts during spaceflight. During Scott Kelly's year-long mission, more than 8,600 genes were significantly altered. Any of these genes, or any permutations of this list, could potentially confer radiation resistance, ranging from DNA repair, free-radical removal, and DNA packaging mechanisms and pathways.

However, as the NASA Twins Study showed, there were also many pathways active during spaceflight that were related to telomeres, including their length, packaging, and maintenance. While telomere shortening is associated with older age, it is highly variable between individuals, and it is not yet clear if modifying these genes would confer additional protection or result in something worse—like cancer. Extensive testing of these genes would take place from 2021 to 2040 (Phase 2), enabling the selection of top candidates to be added as "genome protectors" across multiple generations in Phase 3 (from 2041 to 2100).

Seminal studies by David Sinclair, among others, have further identified genes which can confer higher longevity, such as Sirtuin1 (SIRT1).
Simply increasing the expression of SIRT1 within all cells in the body
would likely have negative effects, especially within subsets of T-helper
cells, where it has been shown to impair differentiation. As such, if
this gene was chosen as an engineering target, it would likely need to
have cell-type specific, controlled expression. SIRT1, like the previously
discussed gene CCR5, is pleiotropic and, as such, will require highly
specific tissue-targeting "tropism" to enable varying levels of expression in a cell-type specific manner. Tropism is direction, or guidance, of
biological material in response to an external guidance—essentially cellular GPS. For example, certain eye therapies (as discussed above) only
work on retinal cells, and certain viruses seem to infect particular areas
of the body, such as HPV that targets the skin or the cervix. Further,
studies conducted by Dr. Min Yu have shown that circulating tumor
cells, which are cancer cells that have entered the blood system, can
exhibit metastatic tropism, which results in a cancer metastasizing to
specific parts of the body based on their gene expression profiles. Engineering this tropism has been historically difficult, but the assumption
is that at the end of Phase 3 (2040), this will be better understood and
leveraged for highly tropic therapies.

Newer clinical approaches could leverage more precise tropism maps
and cell-type specific transcription factors, such that the genetic construct can be incorporated into all cells, but only expressed within the
desired cell types. The actual tropism for these constructs will further
be elucidated through the continual study of viral and cancer tropism,
as well as performing more single-cell expression analyses across all cell
types of the human body to identify cell-type specific membrane proteins, which could be engaged to allow for specific engineering. Some
of this work is already underway with the Human Cell Atlas.

Thus, the eyes, melanocytes, and theoretically any other specific cell
type of a future astronaut could be modified and reengineered to provide an entirely new level of protection. As an example, we could further increase the expression of a gene called MC1R within melanocytes
to help control free radicals from radiation which can wreak havoc on
a cell. This, in turn, could modify the expression of TP53 and Dsup,

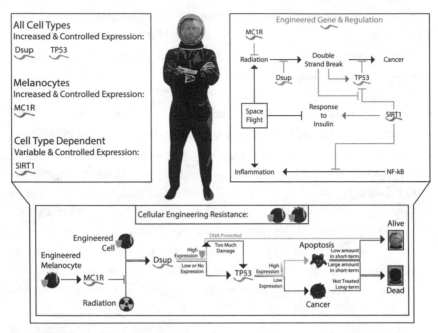

5.3 A genetic framework for radiation protection: By leveraging currently known biological pathways for radiation resistance (black pathways), as well as spaceflight-association pathways that have changed, we can build newly engineered biological networks that can improve the response to radiation and the ability of humans to survive in more harsh environments.

ensuring the proper cellular response. This engineered system, which only has three genes, has multiple layers of protection. To use a soccer analogy: MC1R acts as the defense, controlling for free radicals on the skin; Dsup acts as the goalie, protecting the DNA from any radiation; and, finally, TP53 acts as the referee, making the final decision for how the cell should respond, be it DNA repair or apoptosis. To put it another way, the MC1R and Dsup both act to prevent double strand breaks, whereas TP53 helps guide a decision if the cell should die or not, based on the amount of double strand breaks that got into the net.

So how will we actually target a specific cell type, like melanocytes? To answer this, Shaoqin Gong and Krishanu Saha (at the University of Wisconsin–Madison) developed a new kind of CRISPR delivery system using "nanocapsules" in 2019. These are basically tiny synthetic capsules that can be engineered to serve as the human body's own postal service for genome modifications, dropping off packages to specific cells or tissues. Specific peptides can be added to the surface of these capsules, which acts as the shipping address, guiding the nanocapsules to cell types of interest and attaching when they find their ligands. Even though nanocapsules are small—only 25 nm in diameter—they still have enough room for a tiny molecular package. Gong and Saha showed that they could place a CRISPR-Cas9–enzyme package as well as a gRNA into the capsules.

Gong and Saha further tested several crosslinking molecules which have the ability to hold the polymer together while in the blood, but quickly decompose inside a cell to release the gene-editing cargo. In vitro, human cells happily gobbled up the gene-editing nanocapsules resulting in highly efficient editing in about 80 percent of cells, with little signs of toxicity. Through injecting these nanocapsules into mouse models, they were further able to show proper tropism, targeting and editing of both retina cells and skeletal muscle. Further, the gene-editing nanocapsules also retained their potency after they were freeze-dried and reconstituted, which will be important for the scalability and adaptability of this device as an off-the-shelf product for patients. These results show the feasibility of the highly specific, patient-tailored, in vivo genetic engineering for specific cells.

EPIGENETIC OPTIONS FOR RADIATION DEFENSE

But even if we engineer what we believe to be the perfect radiation-sensing and radiation-resisting human cell through the incorporation of specific gene networks, it might still face other challenges. First, the engineered genes could be "turned off" by the myriad epigenetic mechanisms described above, such as chromatin/histone modifications, and adjusted by DNA base modifications (methylcytosine). Second, they might "drift," so that over time they essentially decay and stop working. To combat both of these issues, one could simply keep editing and modulating the regulation of genes.

Scientists such as Jonathan Weissman and Fyodor Urnov are already putting the core of this idea into play. In 2018, the Defense Advanced

Research Projects Agency (DARPA) solicited ideas for a project called PReemptive Expression of Protective Alleles and Response Elements (PREPARE). The idea is that we can avoid acute radiation syndrome (ARS) by using epigenetic-editing methods to activate genes before radiation occurs. The benefits of this could help astronauts on long-duration missions, soldiers being deployed into an irradiated spot (such as after nuclear warfare), and cancer patients enduring radiotherapy. In 2019, the University of California (UC), San Francisco; UC Berkeley; and the Innovative Genomics Institute received a $10 million award from DARPA to pursue this project.

As a first step toward this goal, Weissman and Urnov are using intestine organoids, which are essentially balls of cells that better mimic the natural 3D structure of a tissue than would exist in a single monolayer of cells in a petri dish. They then screen for genes that protect against radiation when they are turned on or off by CRISPR-Cas9. This iterative process of selectively turning genes on and off allows for a fast and direct way to screen for functional elements of interest, which can then be further enhanced and selected for, as directed evolution.

As humans begin to live long-term on Mars, a unique opportunity emerges to understand evolution in an entirely new light. All our knowledge of evolutionary selection, drift, and sweeps of alleles is based on observations from just one planet. When humans begin living for multiple generations on Mars, we may begin to observe physiological, organ-based, or cellular changes that indicate how the body is adapting to the new gravitational, radiational, and physical pressures. In response, the body, or parts of the body, may create new features that we can observe, model, and utilize for subsequent generations.

HUMANITY'S FIRST BACKUP

Establishing a second planet on which humans and the metaspecies can independently survive will double our chances of long-term survival (from planet-wide risks). However, this only prolongs our time within this solar system. Once we move beyond the solar system, we then increase our chances of long-term survival. We are overdue for a planetary-wide risk that threatens not only us but all life on Earth (e.g.,

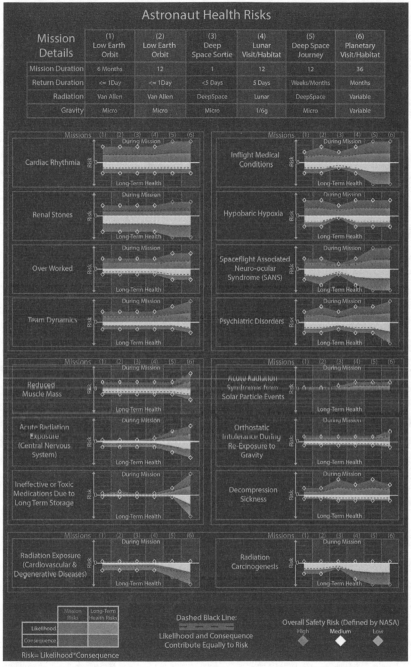

5.4 Risk factors for astronauts: Current risks are stratified by high, medium with and without mitigation, and low, based on mission details, such as where the astronauts will be and for how long. X-axes represent defined missions (1–6) where Y-axis shows relative risk for the mission or long-term health, based on likelihood (lighter) and consequence (darker). (See color plate 9.)

an asteroid). We absolutely need to leave this solar system within 5 billion years, or else all life will be engulfed by the singular most important celestial body that allowed it to exist—the sun. There is no reason that humanity cannot devote itself to survival; indeed, for anything in this solar system to persist, at all, humanity must devote itself to this task, and genetic protections must be identified, tested, and integrated. But before we get to that point, we have to think about new ways to protect astronauts and ensure they can survive.

Of all the risks specified by NASA for future missions, which are measured by green (solved), yellow (low risk), orange (high risk), and red (not yet solved), the radiation risk is still the largest factor for future astronauts (figure 5.4; plate 9). Mars is ideal because of its proximity, similarity of diurnal cycles to those of Earth, presence of water, and high potential for "terraforming." As the first step on a long ladder of species preservation, we must begin to engineer organisms to survive the Martian environment. The first sites of selection and evolution could emerge by 2100, and would serve as a new substrate for targeted evolution. Once this era is completed, in the last half of the "century of biology" and onward, all the genes, cells, and even potentially organs of any organism can become a component of a human cell. By the dawn of 2100, we can start to use all the lessons of evolution on Earth to survive beyond Earth.

6

PHASE 4: PREPARING HUMANS FOR SPACE (2101–2150)

The purpose of models is not to fit the data but to sharpen the questions.

—Samuel Karlin

In Phase 4, we begin to push the limits of the human genome to improve safety in new environments and create a new kind of liberty: *cellular liberty*. In such a time, the genome is no longer fate. The DNA or cells that a person is born with will no longer limit their abilities, who they want to be, or where they want to live. The first step in this process is finalizing the "protected" human genome, regions which need to be preserved for function and viability no matter the setting or environment. This will further ensure our cellular modifications and enhancements are effective and safe, especially before we begin sending humans farther from Earth. This will include the possible addition of new organelles (e.g., modified mitochondria and chloroplasts to harvest energy), new engineered microbiomes on the walls of the space stations, and chimeric cell types. International competition, as well as private enterprise competition, will likely accelerate these innovations.

By 2150, there should be a permanent base of people on the moon, and on Mars we'll begin to build a more permanent station. In this era, human cells will become more complicated, modified, and blended between species, with cells and entire organisms becoming hybrids. In

some ways, this is not surprising, given we are already using these technologies as medical therapies in the early twenty-first century.

HYBRID CELLS

In a large, unassuming concrete building in Hangzhou, China, there is
ongoing research on some of the most cutting-edge science in regenerative medicine and immuno-oncology (IO). New methods for IO are
being utilized to create strange chimeric cells, which are then injected
into patients. These modified cells carry something called chimeric
antigen receptors (CARs), which are modified receptor proteins engineered to be expressed on the surface of cells (usually T cells) to enable
new functions. If the T cell is a weapon that can only work in the day-
light, the CAR suddenly lets the T cell see in the dark, finding previously unseen targets to destroy.

The field of IO and the use of "living treatments" such as CAR-T
have rapidly changed how we approach patient care, amid a broader
shift toward precision medicine and faster deployment of new methods in health care. For example, the United States launched the Precision Medicine Initiative in 2016, with a goal to ensure patient-specific
treatments for as many diseases as possible. In 2017, the Chinese State
Council issued new guidelines to promote the development of "Health-
care Big Data," specifically to create a unified and interconnected
public-health information platform for the entire country, including a
focus on CAR-T therapies.

But T cells are not the only players in the immune system, and they
are not the only ones that can be engineered. The immune system is
composed of many types of cells that are divided into several different
groups. First, there are myeloid cells, which become red blood cells,
granulocytes, monocytes, platelets, and phagocytes, including den-
dritic cells, macrophages, and neutrophils. Second, there are lymphocytes, which can become B lymphocytes (B cells), T lymphocytes (T
cells), and natural killer (NK) cells. While the myeloid fraction, along
with NK cells, composes the "innate" immune system, the B and T cells
compose the "adaptive" part of the human immune system.

B cells are matured like whiskey in the germinal centers of the lymph nodes in the body, where they bask in the body's innate oak casks, making customized antibodies to fight a current (or prepare for a future) infection. T cells, once called the "Holy Grail of Immunology," are defined by the presence of a T-cell receptor (TCR) on its surface, first discovered by Tak Wah Mak in 1984. The cells first start in the bone marrow and develop into several distinct types of T cells once they have migrated to the thymus gland. There, the T cells can be "trained" to use their TCR to detect specific molecules, which are usually presented by the major histocompatibility complex (MHC) on the surface of other cells. Adaptive T and B cells engage with each other and innate NK cells to help identify and destroy potential threats. Together, they create a powerful defensive team for the body.

For human-engineered CAR-T cells, the CAR is matched to an antigen expected to be on the surface of the target cell type, such as CD19 (CD stands for cluster of differentiation, which can define cell types). Thus, CAR-T therapies leverage the combined forces of the body's immune system and human-directed genetic engineering to fight disease, primarily cancer. Whereas naturally occurring T cells almost exclusively target peptides presented by the MHC, and are thus restricted to cells with functional MHC peptide presentation, CAR-Ts can target both MHC presentations and virtually anything else on the surface of a cell.

However, making them is not easy. Customized and engineered immune cells need to be produced in a laboratory under very specific and tightly controlled conditions. Humidity, temperature, growth factors, and subtle molecular differences can lead to either a catastrophic death of the cells or a pure, active population of engineered cells. These cells need to be extracted from either the patient (autologous) or a donor (allogenic), engineered, expanded, and selected, all while conducting quality control to ensure there is no contamination. The final product needs to go through additional quality control, including confirming lack of pathogens, function, and general viability, before finally being infused into the patient.

The first CAR-T used in the clinic, appropriately called "first generation," contained just one intracellular signaling molecule called CD3ζ.

The two main options for modifying cells include changing proteins that sit inside the membrane of the T cell (intracellular) and/or outside the cell membrane (extracellular). The first-generation CD19-CAR-T and CD20-CAR-T cells were used for the treatment of recurrent lymphoma in 2010 and showed a lack of persistence, with CAR-Ts not detected in circulation past seven days. These therapies showed no clear toxicities, though this was primarily due to their lack of persistence and overall lack of anti-tumor effects. Then second-generation CARs were built with a "co-stimulating molecule" (e.g., CD28 or CD137), making them more similar to normal TCRs. These second-generation CARs, first tested in B-cell lymphoma, showed drastically improved persistence, antitumor effects, and production of IL-2. Next emerged the third-generation CAR-Ts, which doubled down on this idea by including two co-stimulatory molecules, further improving the overall control of these cells in vivo.

By 2016, the fourth-generation CARs were built to modulate cytokine and antibody production, with the goal of recruiting more of the endogenous immune system already present in the patient. My lab's work at Weill Cornell in 2020 showed that the response, persistence, and even metabolism of CAR-T cells are dependent upon their stimulatory domains and design. Also, the ratio of the T-cell subtypes used to generate the engineered T cells may play a role in therapeutic potential and cytokine production, with increased T-cell persistence correlating with higher CD8/CD4 ratios. These therapies are powerful, and as a result, complications have arisen throughout these generations, including death, neurological issues, multi-organ toxicities, "cytokine storms," and cytokine release syndrome (CRS).

But all these difficulties have not dampened the enthusiasm for these therapies. By 2020, there were >500 clinical trials using CARs around the world. More CAR clinical trials were initiated in 2016–2018 than in all previous years combined—with the majority in China—but updates continuously emerged in terms of their scope, targets, and technologies. Notably, the United States and China together created >80 percent of all CAR clinical trials posted in clinicaltrials.gov; the site's data were extracted and analyzed to enable patients to find their own trials (carglobaltrials.com).

But ideas for the trials were only the beginning. Even when cells from a human leukocyte antigen (HLA)–matched donor are used to engineer a patient's therapy, the manufacturing process is still delicate and extremely time consuming. Various tweaks to CAR cells are being developed, including the generation of "universal" CAR cells, which are built from cells provided by healthy donors. If CAR cells could be created that are "off-the-shelf," similar to current antibody therapies, then they could be rapidly and widely adapted for many diseases and target combinations.

One company (Cellectis) has already shown that this is possible. The idea of having a stable bank of off-the-shelf cellular therapies derived from nonmatched donors is compelling, but the differing HLAs from donors and patients can lead to inflammation, CRS, rejection, and painful complications from graft-versus-host disease (GVHD). To get around these issues, Cellectis designed edited CD52 and TCR edited T cells which decrease the chance of GVHD while allowing for the selection of edited cells. These cells were further engineered to target CD19 using HLA-matched donor cells and were capable of safely treating two patients with relapsed refractory B-ALL in 2020, who had previously failed all other therapies, sending them into remission.

Numerous other cell-engineering approaches are also possible, such as using nanotechnology to create CAR-T cells inside the body and developing "logical" CAR-T cells with "on" and "off" switches, as well as "and/or" and "and/not" switches, to prevent or limit side effects while enabling the specific targetability of even more cell types. Trials using CRISPR-Cas9 and variations of these gene-editing methods are also being employed to more precisely engineer the T cells and apply them to more cancers and allow them to operate in microenvironments that they would otherwise have no chance to navigate. In 2015, only a handful of cancer types had been tested for CAR-T therapies. By the end of 2020, dozens of cancers (both solid and liquid) were being actively treated and tested in clinical trials.

Beyond CARs, there are two other kinds of adoptive cell transfer (ACT) therapies that use immune cells from patients to treat their own disease. These include tumor-infiltrating lymphocytes (TIL) and engineered cells with new TCRs. TILs are active lymphocytes that have penetrated a tumor or its nearby environment, and they can be engineered to become more active and have shown promise in melanoma and cervical cancer. The other ACT therapy is to engineer T cells with TCRs from other cells. TCRs which are identified to have potent antitumor effects within one patient can be adapted to other, HLA-typed patients through the integration of these sequences into the patient's own cells. However, given that the majority of TCRs engage with MHC presentation, they are less applicable in a truly universal setting like CARs. Some efficacy has been seen with TCR-based therapies, with an overall safer toxicity profile than CAR-19 cells. However, the ACT therapies that had advanced the furthest in clinical development by 2020 were CARs.

Indeed, by 2021, tens of thousands of people were walking around with engineered, hybrid cells inside of them. These therapies resulted in fundamental shifts in how we treat disease and our own understanding of the plasticity of life. Biology is not passive and impermeable. Instead, it has the potential to be engineered; biology is now capable of consciously directing the creation of new life, within minutes, in ways that evolution may never have had the time to produce. This capability reaches far beyond immune cells—to any cell in any species that exists or will be created. Some hybrids exist on Earth now that can guide what we can build for new planets.

HYBRID GENES BETWEEN SPECIES

Floating in the waters around Boston and New York is a strange, small green-hued hybrid sea slug: *Elysia chlorotica*. This unique species has the ability to become plantlike by stealing fully functional, photosynthesizing chloroplasts from the algae it eats—a process called "kleptoplasty," which literally means "theft" of the plasmids (chloroplasts) or organelles. While absorbing DNA and moving mobile genetic plasmids in bacteria is common, moving entire systems is rare in larger organisms.

Sometimes called "solar-powered sea slugs," the Elysian species uses the green chloroplast as camouflage against predators. Algae normally have a hard, thick plant-cell wall that prevents any breakage or invading species—so how do Elysians get the chloroplasts into their body?

With straws, of course! Elysians have built-in, molecular straws that enable them to pierce through the algae's walls and suck out the chloroplastic goodness, turning bright green. If the slugs don't eat enough of their "vegetables" (chloroplasts), they become brown with red pigment spots.

Surprisingly, the chloroplasts can survive for months, or even years, within the large branching digestive system of the (now green) sea slug. Much like the phagocytes of the human immune system, the Elysian phagocytes can engulf the algae easily, and then integrate the chloroplasts into their own biological systems. Even when embedded into their bodies, the chloroplasts still function, capture sunlight, create sugar, and exhale oxygen. Although it was first thought that the eerie green sea slugs needed the chloroplasts to survive, it turned out they did just fine without the light. A researcher named Sven Gould showed that even without light, the slugs' survival and weight were about the same. So, this is to some degree a recreational feature of the slugs, as if their favorite way to spend their day is the cellular thieving of green, internal adornments.

But these little green thieves raise the question—how do the chloroplasts survive and function within their bodies? In normal plants, chloroplasts require 90 percent of their essential proteins to come from the nucleus of the plant host. They are basically mooching roommates. Clearly, these sea slugs can also accommodate the needs of their visiting chloroplasts, but how? When looking for possible genes that could support chloroplast survival and photosynthesis, James Manharte and other researchers noticed that a key algal gene, psbO, was in the sea slug's DNA. psbO is a critical gene because it encodes for a manganese-stabilizing protein which is part of the photosystem II complex of the chloroplast.

Most importantly, the DNA sequence of the sea-slug gene and the algal gene were almost identical. It seems as if the sea slug had, long ago, borrowed the gene from the algae and never returned it. This opened up the exciting possibility of horizontal gene transfer (HGT), where a gene from an organism is "horizontally" moved from one species to another. This is in contrast to "vertical" gene transfer, where DNA moves between one generation and the next.

But how could these researchers be sure it was HGT? Initial evidence showed that the gene was already present in the eggs and sex cells of *Elysia chlorotica*. However, the genes did not seem to be active when subsequent work examined their RNA, with further analyses in 2017 indicating that there was actually little evidence of these genes in the egg (germline) DNA. Thus, while the mechanism of how chloroplasts captured by *Elysia chlorotica* can survive so long is still somewhat of a mystery, it is clearly possible, and it might have been helped by HGT.

Another example of HGT comes from tardigrades, which are the famed "water bears" that can survive in the vacuum of space (also featured in chapter 4). Dozens of tardigrades' genes are likely to be derived from HGT and may also contribute to the biology of the organism. This process of "fluid genes" between species is a key driver of evolution because millions of years of selection pressure in one context can suddenly be positioned in an entirely new context for a new genetic enrichment of features and functions.

CHLOROHUMANS THE SIZE OF TWO TENNIS COURTS

Could humans mimic our thieving friends, *Elysia chlorotica*, and photosynthesize instead of always having to eat with our mouths? To get chloroplasts to work in humans, we would have to make some big assumptions. The first assumption is that human skin cells would be capable of supporting the chloroplasts. This support would require our immune system to not reject them and that melanin (the pigment that gives skin its color) would not interfere with chloroplasts' functions. Beyond this, the chloroplasts would need to survive and be functional, but the *Elysia chlorotica* system shows it is possible.

The next assumption we have to make concerns the chloroplast photon capture efficiency within its new, human host. No chemical reaction is ever 100 percent efficient, mostly because of the second law of thermodynamics, biophysical limits of efficiency, and other errors. So, what percent of the sun's energy should we assume the new "green human" can capture? Estimates suggest that plant efficiency of capturing photon is only about 5 percent. So, we will assume that the new "chloroskin" cells would act similarly.

The next question is about how much energy we could get from our chloroskin. On average, each human has about 1.7 square meters of skin, but even if fully naked, only half of that skin would likely be exposed to the sun (e.g., when you're lying on your stomach). On a bright day, sunlight energy levels are about 300 watts per square meter, which is enough to power a normal light bulb for about three hours. Finally, to be conservative, assume the photosynthesis biochemistry inside of the chloroplast is only 75 percent efficient. Given that input, the chloroskin would only collect about 34 kilojoules of energy per hour. An average-sized human needs about 10 million joules per day to survive.

Thus, for a human to function at normal energy levels, 290 hours of midday sunlight would be needed to collect enough energy to get through one day. However, to reach the required energy, more skin could do the trick. If a human epidermis was expanded 300-fold (1.7 m^2 × 300), which is about the size of two tennis courts, a chlorohuman lying on their stomach would only need to sit in the sun for about one hour. Therefore, a chlorohuman could go on a lunch break, unfurl their newfound skin in a large empty field somewhere, get a meal while maybe taking a nap, and then close up their skin and head back inside, full and satiated.

MOBILE GENES AND SEMIGENES

Given that chloroplastic kleptomaniacs exist in the animal kingdom, it should come as no surprise that there are other small, mobile molecules moving between species too. In 2010, Alain Robichon discovered high levels of carotenoids in aphids, small insects that can be found in leaves around the world. This, on its own, isn't too strange, considering animals need carotenoids for a variety of cellular functions including vision, coloration, and vitamin processing. The peculiarity comes from previous research by Nancy Moran and Tyler Jarvik, which showed carotenoids were not present in the diet of aphids. The orange and red organic pigments, which give the characteristic autumn colors to pumpkins and tomatoes, were thought to be made only by plants, algae, bacteria, and fungi—yet here is an insect that apparently could make them on its own.

Robichon and his team set out to discover what these small insects could possibly be doing with such high levels of, apparently, synthesized or stolen carotenoid. They first noticed that cells with high levels of carotenoids also had elevated levels of adenosine triphosphate (ATP)—essentially the gasoline of the cells. They then noticed that the levels of ATP would change depending on the insect's exposure to light. Place the insect in light, ATP goes up; put it in the dark, ATP goes down. To further test their response to light, the team split the aphids into two teams: those with high levels of carotenoids and those with lower levels. As expected, the group with higher carotenoids was able to absorb more light. The team further showed that the carotenoids in the aphids were close to the surface (0–40 nm), exactly what one would expect if the carotenoids were being used to capture sunlight.

Then, in 2012, Moran and Jarvik completed a phylogenetic analysis that identified genes in the insects which were almost identical to those in the carotenoid pathway in fungi. They looked at thirty-four aphid species around the world and noted that all aphids had at least one copy of this gene (lycopene cyclase/phytoene synthase), and some aphid species even had seven. In contrast, all fungal genomes only have a single copy. The closest-living relatives of aphids, called adelgids, also showed evidence of possessing this pathway. Thus, given enough time, the genes from one entire kingdom of life can move into another and further provide entirely new functions.

Importantly, these are not the only examples of genes moving from one organism to another. HGT has been shown in bacteria to fungi (*Saccharomyces cerevisiae*), bacteria to plants (*Agrobacterium*), bacteria to insects (*Wolbachia*, in beetles and bed bugs), organelle to organelle (in parasites of the *Rafflesiaceae*), plant to plant (hornwort to ferns), fungi to insects (as above with pea aphids), human to parasite (*Plasmodium vivax*), virus to plant (tobacco mosaic virus), and possibly plants to animals (Elysia species, above). The most scandalous and extensive HGT that has ever been observed is a jump from bacteria to animals in small mites that live throughout the world's oceans, bdelloid rotifers. An estimated 8 percent of their genes are derived from bacteria.

However, the most striking example of genes moving is when they do so en masse. The endosymbiosis theory of mitochondria and chloroplasts holds that, at some point, these "minibacteria" were ingested or merged with eukaryotic cells. Rather than die or split up, they decided to marry—and have been together in loving cellular matrimony ever since. This enabled not just one gene to get moved, but entire networks, membranes, and new abilities. For example, the ATP in human cells, on which human biology depends for sustenance and existence, is not even made by the human component of a cell; it is created in the mitochondria.

Notably, the transfer of genes from the mitochondria to the human genome, and vice versa, is a process that is still ongoing. Nuclear mitochondrial DNA segments (NUMTs) are a result of this engagement and exist where the mitochondrial genes have migrated, like nomads, from the mitochondria to the human nucleus. The DNA in our cells works regardless of where it came from, meaning that our gene networks do not choose their place in our cells with any regard to their history; rather, their place is determined based on what is needed. This same principle that applies to life on Earth can be readily applied beyond Earth as well.

Given these widespread and pervasive examples of exchange of DNA between species, it is not unexpected, or even unnatural, to begin to think about doing so in human cells. Because our own human lineage only provides evolutionary lessons from the past few million years, we would be better served by taking the lessons from billions of years of evolution for us to survive on faraway worlds.

ONE CELL BECOMING ANY CELL

While moving genes from one species to another is exciting, what is perhaps even more thrilling is the act of turning one cell into any other cell. Given that every cell is preloaded with the knowledge to function as any cell that makes up the organism, it should be possible to convert one into another, given the right genetic and epigenetic tools to press the right buttons. This could be anything from a blood cell to a skin cell, or even to a totipotent cellular state, which can then eventually develop into a new child.

There is a rich body of work around stem cells, induced pluripotent stem cells (iPSCs), and cellular reprogramming. iPSCs are usually created from adult somatic cells, such as from the skin or blood, which are then bathed in a special cocktail of transcription factors and nutrients. Because they are very similar to embryonic stem cells (ESCs), iPSCs have inspired entire communities of clinicians and researchers to dream big. The California legislature was so excited by the promise of ESCs that it launched its own funding initiative for stem-cell research, to rebuff the US federal ban, in 2005 (Proposition 71).

However, you cannot simply sweet talk a cell into doing anything you want; specific molecular coaxing is required. This push of a differentiated cell (terminal state) into a pluripotent state was first shown in 2006 by Shinya Yamanaka and John Gurdon using mouse fibroblasts and four specific transcription factors: Sox2, Oct4, Klf4 and c-Myc—now commonly known simply as the "Yamanaka factors." Human iPSCs then came in 2007, derived from skin fibroblasts. This finding was gigantic, not only because it helped us understand the basics of cellular differentiation, but because it obviated the need for sacrificial embryos, which had widespread public policy implications. Both ESCs and iPSCs showed pluripotency, rapid self-renewal, and similar gene-expression patterns.

Within six years of the US ban on federal funding for ESC research, there was a seemingly new way around it. However, iPSCs are not identical to ESCs and are still in their early years of research. One of the biggest worries of iPSCs is tumor formation. If either iPSCs or ESCs are directly injected into a patient, there are risks of teratoma formation. In the world of tumors, teratomas are essentially gremlin children—composed of all three developmental layers (ectoderm, endoderm, and mesoderm); some result in extremely odd formations such as growing hair and teeth. In the case of iPSCs, there is an increased worry of tumors due to the forced overexpression of known oncogenes and the artificial selection pressure during their creation. Further, given that these are derived from adult tissues, there is always the chance that the cell has acquired additional mutations (like in clonal hematopoiesis) that may be oncogenic. Also, experiments have shown distinct differences within both the epigenetic and transcriptional landscapes of these cells. However, even after many protocols and years of experimentation based on ESCs, iPSCs have not completely replaced ESCs in research or therapies.

Nonetheless, iPSCs offer many advantages which ESCs simply cannot. Given that ESCs are formed from terminated embryos, and iPSCs are derived from an individual's somatic cells, iPSCs can essentially act as a person's own, easy-to-create surrogate twin. To create these replicas, blood can be extracted from a patient, turned into iPSCs, and then differentiated into a host of various cell types that can act as models for the patient's actual tissue. As an example, a patient with familial amyotrophic lateral sclerosis (ALS) (around 5–10 percent of cases) could have their fibroblasts converted into motor neurons, the disease cell type, which can then be used as a drug screen to identify potential therapies for the patient. In larger cohorts, this method could be used to elucidate the disease mechanism of action and novel therapies. Essentially, any genetic disease that affects a single cell type or tissue could be solved through this process.

Both human ESCs and iPSCs are pluripotent, not totipotent. By the strictest definition, a totipotent cell has the ability to be placed into a zona pellucida (the initial protective and nutrient-rich shield of an oocyte), which can then be placed into a mother-to-be (or exowomb) and fully develop into a functional offspring. Pluripotent cells are cells that are capable of giving rise to all tissues that make up an adult organism. What they cannot do is create all of the early-forming developmental layers required to support the fetus, such as the trophectoderm (which becomes the placenta).

There are a number of reasons why we have yet to establish these lines, including when the ESC is extracted from the embryo, how the cells are extracted, the culture conditions, and even a philosophical reason of "personness." For approximately the first two weeks of an embryo's development, any cell in the embryo can be "popped out" to create a perfect clone. Any cell, some of the cells, or half the cells could be separated to create a twin; this is often how identical twins or triplets are made (e.g., Mark and Scott Kelly). However, around day fourteen, a "primitive streak" emerges, creating left/right, top/bottom, and front/back axes for the embryo. Once this streak appears, as a differentiated line down the middle of the embryo, it is just a single embryo. It can no longer become hundreds of people; it has become a single person. This is why most ESCs are not grown past fourteen days.

Moreover, working with ESCs can be very scientifically challenging, as they will often spontaneously differentiate when they get lonely; they are very temperamental while in cellular culture conditions. There is a lot of ongoing work on new ways to form totipotent cells, including further engineering through silencing differentiation signals and altering culturing conditions (as described above with Ali Brivanlou's work on embryos). Indeed, our ability to turn any nucleated cell into any other cell is simply a matter of solving for what buttons need to be pressed. If the full epigenetic code were known, and if we could modify any DNA locus, RNA molecule, and histone state, in theory, we could convert any cell into any other cell, at any time. The only real challenge would then be in maintaining that cell state. The field is young, and over time, these questions will eventually be answered.

COMPLETE PARENTAL LIBERTY

The idea of total cellular flexibility has some precedent. This is an old idea, called parthenogenesis, which is a method of asexual reproduction whereby embryos can be created without fertilization, but, at least in the twenty-first century, this ability is not present in mammals. As is the case with any biological function, however, it could be engineered as long as we have a template on which to build. In principle, you could go beyond one-person or two-person babies to a three-person or more baby, or even further enable two females or two males to have their own genetic child. These possibilities are all just a matter of engineering.

In 2004, two female mice became the first mammals ever to mother an offspring using only their maternal genetic and cellular material— what some called the first true "lesbian mice"—in an experiment by Tomohiro Kono. Historically, this seemed impossible, due to parent-specific epigenetic modifications (called imprinting) during the creation of egg and sperm. The imprinting process ensures that there is a sex-specific expression of genes, an "epigenetic imprint," based on whether the gene came from the father or mother. This imprinting has been shown to persist even beyond one generation, based on work by Oded Rechavi and others. However, until the 2004 paper by Kono, it was unknown if genomic imprinting would be too great a barrier to parthenogenetic development.

Kono and his team created an oocyte with haploid (containing only one set of chromosomes instead of the usual two) sets of two maternal mouse genomes, resulting in the creation of their offspring through parthenogenesis. Yet some assembly and tweaking was required. The team modified the expression of two key regulatory genes (Igf2 and H19), as well as some other imprinted genes through the creation of a large (13 kilobase) deletion in H19. The two mothers eventually became grandmothers when their offspring went on to have its own baby. This groundbreaking work was proof that males are not required to produce viable offspring (at least in mice).

But can males create progeny by themselves as well? Indeed, they can. In 2018, Wei Li, Qi Zhou, and Bao-Yang Hu showed that males, too, can have a two-male genetic offspring—though these "two-male mice" did still need females to carry and give birth to them (yet another use case for exowombs). Li and his team created their necessary cells through the use of haploid embryonic stem cells (haESCs), which are made from either a purified egg or sperm—in this case, obviously sperm. Again, the challenge of imprinting persisted, so the team used CRISPR-Cas9 to delete regions of the genome involved in imprinting. They reproduced the 2004 work, observing that female-only babies required only three deletions, but males were more complicated (insert your own joke here), requiring seven deletions. Also, for the males, an egg was still needed to start the embryogenesis, and so an egg donor (technically a three-parent baby due to the mitochondrial DNA) was used. Li and colleagues then injected the haESCs differentially into sperm and an immature egg stripped of its nucleus. The offspring were again viable, thanks to the well-planned engineering.

This means that two human parents from either sex (male or female), someday, could have children as they see fit. Further, it's not hard to envision a process by which our iPSCs, every cell to any cell, could be combined with additional engineering to direct female iPSCs into sperm or male iPSCs into eggs, both of which could be fully developed within an exowomb and require no donors or minimal long-term genetic alterations.

Ethically, this represents near-complete parental liberty and "cellular liberty." If sex begins to be decoupled from reproduction, then more autonomy is granted and the deontogenic ethic is supported. With artificial uteruses and exowombs, the process of creating a child can be done alone or in many other combinations. Given all that is possible in 2021, much more will be possible by 2150, including having a child by yourself without ever actually having to give birth, same-sex couple children, opposite-sex couple children, three-parent babies, and perhaps even "build your own" children from more than three people with synthesized segments of their genomes combined. This opens a new era of completely free cellular, reproductive, and parental liberty created by biological engineering.

ENGINEERING HUMANS FOR SPACE

As such, by 2150, improvements made to cellular and genetic engineering will result in less of a distinction between germline and somatic cells. By then, adult tissues can undergo somatic engineering just as readily as zygotic germline editing, and it is likely that a significant proportion of people in the United States will be zygotically edited or will be the product of someone who is. Given the ability to directly remove genetic disease before a child is born, and thus to remove the disease from their entire lineage from that point on, fewer diseases will require somatic engineering interventions. The remaining medical uses of genetic engineering will primarily be to correct environmentally driven conditions (including cancer and infections) or age-related genetic or epigenetic drifts. The overall usage of somatic engineering will not necessarily decrease over this period, but will instead shift toward more elective procedures such as sex reassignment, aesthetics, or even temporary epigenetic changes. Thus, radiation-resistant editing for astronauts can occur at the adult or zygotic stage. Medical uses of genetic engineering will be primarily for environmental conditions including cancer and infections.

The majority of biological research will then be focused on "improving" genomes through combining elements from multiple different species or newly synthesized chimeras, as opposed to "fixing" them as was the mindset in the twentieth century. The distinction between temporary and permanent will fade, as any genetic modification could be undone through additional somatic engineering. The only question is whether the specific modification will persist without additional interventions (genetic) or not (epigenetic).

Cellular and genetic engineering will then become so common that high school students may be given genetic design projects as homework, which they can ponder while looking out their window at a moon that twinkles in the sky, full of a moonbase's city lights. Engineering life on Earth is widespread and deeply engrained within not only us, but our pets and other animals. Genetic engineering enables people on Earth to live *how* they want; its next challenge is to enable people to live *where* they want beyond Earth. All design substrates (de-extinctified, extant, and newly created) will then be deployed to help us understand how we can enable humans and other species to better adapt to spaceflight and spacelife.

7

PHASE 5: SYNTHETIC BIOLOGY FOR NEW HOMES (2151–2200)

Essentially, all models are wrong, but some are useful.

—George E. P. Box

By 2151, in vivo genetic editing will likely be commonplace, safe, and, in some cases, even "recreational." Several space stations will be in permanent orbit around Earth, the moon (Lunar Gateway), and Mars (Base Camp). The majority of severe, inherited human diseases will be solved and "edited out." The further development of technologies based on twenty-first-century products like CAR-Ts will enable the modification and subsequent creation not only of novel cellular and immune functions, but of entirely new cell types (borrowed or blended from other species). We will further be able to continually monitor these living therapies, enhancements, and any other cell type in the body, enabling us to detect problems before they manifest. Through the continual research into the modification and integration of genetic elements from one organism into another (e.g., tardigrade Dsup into human cells) that will occur in Phases 3 and 4, we will learn how to optimally adapt naturally occurring elements for new roles, as well as begin to create our own purely synthetic genes and networks.

From this fertile cellular-engineering base, the rise of "occupational" gene enhancements will emerge to improve safety and performance for specific jobs. As an example, astronauts could obtain somatic engineering before leaving Earth, which decreases health risks through improving radio-resistance, insulin response, and cancer prevention/ surveillance, or even resource-saving genetic editing protocols, such as reducing the amount of required oxygen for cellular function or metabolomic intake. Astronauts will be cellularly prepared in mission-specific contexts that depend largely on the duration of the trip and whether they are traveling to harsh environments of the solar system (e.g., Kuiper Belt asteroids, Titan, or other moons of Saturn) or relatively more benign ones (e.g., Mars). But even humans who wish to simply live on already established safe areas outside of Earth will likely be required to undergo a minimum degree of genetic engineering just to relocate. This will be the beginning of the settlement of other planets and the genesis of new, entirely synthetic genomes, but it will also come with the risk of disrupting any life that is already there or fossils from previous epochs.

This raises questions of "planetary protection," which is bidirectional. The first aspect of planetary protection is to avoid "forward contamination," whereby we bring something (accidentally or on purpose) to another planet. This is important to ensure the safety and preservation of any life that might exist elsewhere in the universe, something with which humans have a poor track record (e.g., smallpox on the blankets given to Indigenous people of North America or the rapid spread of SARS-CoV-2 in 2020). Secondary to this is to ensure the genuine discovery of non-Earth life, rather than the false identification of an alien-looking, but actually only Earth-grown, contamination. Microbes could potentially accompany us on a trip to Mars, even after radiation and sterilization procedures, and their genomes may change so much that they look truly otherworldly. If these were then later found in the Martian sand, it could potentially spark misguided research into the universal features of life. The other component of planetary protection is avoiding "backward contamination," whereby something that we bring back to Earth presents a risk to the native terrestrial organisms. This is the theme of many science-fiction movies, where an evil "alien" invader threatens all life on Earth.

Neither of these scenarios is ideal. But the first scenario of forward contamination is actually unavoidable, because all life in the inner solar system will eventually be engulfed by the sun. Deontogenic ethics would require us to engage in some amount of inevitable, forward contamination of other planets in order to save not only Earth's life but as many life-forms on other planets as possible (or at least the knowledge that they once existed). To plan for our visit to these harsh worlds by 2201, we first need to know how spacecrafts are made in 2021.

LIFE-FORMS ALREADY SENT

Amid the arid air of Pasadena, California, at the Jet Propulsion Laboratory (JPL), stand a series of buildings dedicated to the ongoing effort of sending rockets, probes, and robots out into the solar system. Understated, plain signs declare the functions of rooms and buildings, such as "Spacecraft Assembly Facility (SAF)," "Extraterrestrial Materials Simulation Laboratory," and "Mars Rover Testing Area." This is where rovers (e.g., the Mars 2020 rover, Resilience), satellites, and many NASA spacecraft components are designed, assembled, and tested before their eventual launch to go out and explore areas of the universe humans have yet to travel. Humans' spacecraft have already reached farther than our own solar system (defined by the sun's heliopause) in 2012 (Voyager 1), and again in 2018 (Voyager 2). Eventually, the Voyager probes will send back data on nearby solar systems unlike anything we've ever seen before. An ongoing question about these Voyager probes, and really any probes, is: Did anything else go with them?

To test this idea, regions of the JPL campus, most especially the "clean room" where spacecrafts are built, undergo continual monitoring for the presence of spores, bacteria, viruses, and any other life-forms. This is an ongoing effort to quantify and limit any degree of forward contamination. Walking around the construction sites on most days is a scientist named Kasthuri Venkateswaran, a microbiologist who has published more than anyone else about microbes and spaceflight, as well as another pioneering extremophile scientist, David Smith. Both are constantly working to improve our understanding of which organisms can not only survive the trip to space but thrive. This is no easy feat, though, as they are confronted by a "low-biomass collection" problem: there should be an extremely small amount of bacteria or viruses present—potentially even none—but it only takes one cell to expand. So, how can you test for something that is barely there at all? How can you be sure nothing is present? Further, if we miss something, and it makes it to space—what will happen?

ASTRONAUT MICROBES

Some of these questions have already been answered in preliminary studies of the spacecraft, astronaut, and space-station microbiomes by our group and others. Numerous longitudinal studies, including the Twins Study and studies from Hernan Lorenzi, Duane Pierson, Alexander Voorhies, Mark Ott, and colleagues, have shown that microbes don't exactly just sit still while in space. Specifically, Lorenzi showed that bacteria on the skin of various astronauts became more similar over time. This makes sense because there was nowhere else for them to go; they were spending so much time enclosed in the same aluminum tube. This "blending" of microbiomes was partially driven by a drop in the abundance of various species, including a fivefold reduction in *Akkermansia* and *Ruminococcus* during flight, and a threefold drop in *Pseudobutyrivibrio* and *Fusicatenibacter*. As we showed with the Twins Study, most of these changes reverted to preflight levels after the astronauts returned to Earth. Moreover, these changes in the skin microbiome may contribute to the high frequency of skin rashes and overall skin hypersensitivity experienced by astronauts in space.

Potentially more worrisome than the relative change in the ratios of different bacteria are "space zombies." We haven't found any space zombies in the "reanimated-human" sense, only in the reanimated, latent-virus sense. Previous work from Satish Mehta at Johnson Space Center showed that the incidence of herpes reactivation appears higher in space, and viral shedding may end up floating around the ISS. As of 2019, forty-seven out of eighty-nine (53 percent) astronauts who went on short space-shuttle flights (weeks), and fourteen out of twenty-three

(61 percent) astronauts on longer ISS missions (three to six months), showed evidence of herpes viruses in their saliva or urine samples. These astronauts further showed signs of active infection through elevations of cytokines and immune cell markers.

So, if something comes out of your roommate on the ISS, how long will it float around? Also, what exactly is being shed around the ISS? To answer these questions, older studies have traditionally used culturing methods to assay what can grow and survive on the space station. However, these methods can often result in bias-selection, where some bacteria simply don't want to grow in the culture conditions, or other bacteria really, really, enjoy this new environment and outcompete all of their friends. However, since 2015, more and more microbiological studies have deployed "culture-independent" methods for a thorough examination of the changes of bugs in space. These methods employ NGS, as described in chapter 4, to take a given sample, extract all the DNA, and then "shotgun" sequence it. As the term implies, it is like taking a shotgun to the cells of a sample, blasting them up into billions of small DNA fragments, and then sequencing each piece. Each piece (or sequence "read") can then be aligned (or mapped) back to the known genomes of species that are already present in sequence databases. This enables a quantitation of the abundance and type of organisms present in any one sample. One can also perform *de novo* (from nothing) assembly of the short reads into longer reads, much like putting pieces of a jigsaw puzzle back together. This is the best means to examine any organism that does not grow on normal bacterial media; we can even find organisms that have never been annotated before.

NGS methods have already shown that space-based bacteria can have increased virulence, become more resistant to antibiotics, improve biofilm formation, and even get thicker. In 2016, work by Luis Zea, Shawn Levy, and colleagues noted that changes in the types of antimicrobial-resistance (AMR) markers and gene-expression pathways were specifically enriched in response to spaceflight and that they spanned a variety of pathways, including protein synthesis, nucleic-acid binding, and metabolism. As expected for any extreme environment, the ISS exerts a strong selection pressure on bugs.

THE HALF UNKNOWN

Dr. Venkateswaran's analysis of organisms on the ISS showed a large span of variations ranging from what looked like an entirely new species of an existing genus to fragments of DNA that did not align to any known genome at all. However, this is not unusual by itself. When the shotgun-sequencing data from the NYC subway was analyzed by our group in 2015, and again globally in 2021, we showed ~50 percent of the sequenced DNA was completely novel and had never been seen before. Similar numbers were observed when the geneticist Craig Venter used shotgun sequencing in the Sargasso Sea. Thus, the biggest factor driving the mystery DNA found on the ISS is likely not aliens or anything otherworldly at all, but instead limitations to our current reference databases. Given an estimated trillion species on Earth in 2020, with only about 100,000 having reference genomes to which we can compare, it is a small miracle that we can routinely identify anything we capture at all.

As we sequence more and more diverse locations around the world, the number of "unknown" fragments of DNA will decrease. A new species from the ISS was even named after Kate Rubins (*Kineococcus rubinsiae*) in 2020. However, the unknown fraction can never reach zero; there will always be a sequence that we have never seen before. This is because life is always evolving, especially within some bacteria and viruses that replicate in a matter of minutes. Thus, even if one had a complete and total genetic catalog of all life living anywhere on Earth, it would be incomplete only a few minutes later.

However, this does not mean we should stop searching for new unknowns. To even attempt our planetary-protection initiative, we need to analyze all fragments of DNA we can find in any and all spacecrafts (which, it just so happens, is exactly what we do). With the work by Drs. Venkateswaran, Smith, and others at JPL, shotgun sequencing and surveillance of the spacecraft is now a standard metric for the estimates of forward contamination. These data are then compared to the total index of all sequenced DNA that has ever been seen on Earth to identify what is present, and what might be accidentally propelled into space on the next mission.

So far, the results are very clear. We have *unquestionably* contaminated Mars, and possibly other planets as well. While the "bioburden" of these missions has been consistently low, it has *never* been zero, not even the *Viking* landers of the 1970s. For Mars, this is particularly problematic, due to the planetary dust storms capable of picking up and spreading the contamination across its surface. As such, chances are low that we will be able to maintain contamination-free areas of Mars above ground. This means that if any life is found on Mars, it would need to be dramatically different from anything seen on Earth for us to be confident that it is Martian in origin.

In other instances, planetary-protection measures have been far more rigorous. The *Galileo* spacecraft deliberately took a nose dive into Jupiter's devastating atmosphere in 2003, erupting in hellish fires of decontamination. This was done to protect the *Galileo* probe from contaminating its own discovery of possible life-harboring oceans under the surface of the moon Europa. Also, the *Cassini* probe exploring Saturn was purposely burned up in the atmosphere to prevent it from creating any forward-contamination. Burning up a probe in the atmosphere is sometimes the best way to get rid of any contamination. If the probes could speak, they might tell you a different story.

SYNTHETIC BIOLOGY'S ROOTS

So, let's say we find an organism on another planet and we are confident it's not from Earth. What would it look like? The reality is that we don't know, as our current understanding of life is based on just one planet, Earth. Further, the truth is that our own understanding of what defines life on Earth changes as we identify new extremophiles and make radical advances in synthetic biology.

Synthetic biology is a relatively new discipline. Even the (now seemingly obvious) idea that cells are regulated by modular, molecular networks arose in 1961 when François Jacob and Jacques Monod were studying *E. coli*. The ability of bacteria to use lactose (such as in milk) as an energy source is driven by the lac operon, a group of genes with a single promoter, which when utilized energizes the cells. This configuration defined the first component of a genetic circuit.

This genetically controlled "on-and-off" switch led Jacob and Monod to imagine all genetic functions to operate as programmable circuits, like a computer. Then, in the 1970s and '80s, the seeds of synthetic biology began to sprout with the help of molecular cloning. Once restriction enzymes enabled cutting and pasting pieces of DNA together, the ability to modify the genetic code of different organisms, or even combine parts of different organisms together, became possible. Then, in the early 1990s, genomes from multiple organisms were sequenced, including *Haemophilus influenzae*, *Saccharomyces cerevisiae*, and eventually *E. coli*, paving the way for comparative genomics where segments across and between genomes are found, annotated, and analyzed against each other.

The year 2000 marked the dawn of the first synthetic circuits which were created and "transplanted" into another system in an organized and structured fashion. These circuits were integrated into *E. coli* in 2002 and 2003, leading to the first example of engineering bacteria to produce drugs (Artemisinin). These experiments opened the door to an entirely new way to synthesize virtually any small-molecule therapeutic we may ever want or need. However, this process was challenging, slow, and laborious. Almost all molecules require many steps to be synthesized and further augmented, so they can be readily utilized by human cells. As already discussed with many other biological technologies (e.g., codon optimization), this, too, needed innovation to perfect.

To help with this, in 2004, the International Genetically Engineered Machine (iGEM) competitions brought together high school, undergraduate, and graduate students to see who could engineer the best organism. iGEM groups have spearheaded the development of multiple types of standards technologies and programming software for synthetic biology. In particular, the Synthetic Biology Open Language (SBOL) is a framework for designing circuits, cell constructs, and synthetic-biology components that can be shared with other people around the world. Even the first concept for this book (a ten-stage plan for 500 years) was conceived in 2011, as a wiki post for the Mason Lab–Weill Cornell–iGEM team.

Another effort, which started in 2004, with the intent to spark collaboration was the synthetic biology (SynBio 1.0) international meeting, bringing together a myriad of disparate fields—including engineering, genetics, chemistry, physics, electrical engineering, and design—and merging them into one composite field. The true test of systems biology was whether we could modify elements of a cell's genome, or insert sections of another organism's genome into it, and accurately predict what would happen to it. Drew Endy, a biologist at MIT and later Stanford, called these integrated sections "biobricks." Thus was born the Registry of Standard Biological Parts, which is still active today and has led to many hypotheses about how much, and what, can be moved between species.

Many of the ideas that were dreamt up during those first SynBio and Biobricks meetings would be realized shortly thereafter. The first bacteria that were engineered to invade cancer cells were created in 2006. In 2007, the first engineered bacteriophage that could control biofilms was created. Biofuels began showing promise using *E. coli* in 2008, and then oscillators and logical switches also appeared in 2008.

However, these were relatively simple systems primarily implemented in *E. coli*. Clearly, larger genomes had much more complicated molecular regulation (e.g., epigenetics and histone codes), which poses additional engineering challenges. The Gibson DNA assembly method was described for the first time in 2009, which is a bit like combining Lego pieces to form a larger Lego, then combining more pieces with that larger Lego to make even larger Legos. This method coincided with the engineering of an "edge-detector" circuit with the ability to count changes in a system, leading to the idea of "genetic clocks" that could be coupled with population-sensing mechanisms in 2010 and 2011, and even Boolean logic was demonstrated for bacteria in 2011.

MINIMAL NECESSITIES FOR LIFE

Also in 2010, researchers reached a giant milestone in synthetic biology. The first chemically synthesized genome was made from scratch, as described in a paper appropriately titled "Creation of a Bacterial Cell Controlled by a Chemically Synthesized Genome." This work showed that a completely synthetic genome (made from simple nucleotides) could be created, placed in a cell, and "rebooted" to activate life. Just as with exowombs, this was a nail in the coffin of "vitalism," or even "neovitalism." By 2012, entire chromosome arms of yeast were synthesized by Jef Boeke, and by 2013, genetically engineered *E. coli* were created to produce Artemisinin commercially. By 2020, entire companies like Ginkgo Bioworks (led by Drs. Jason Kelly and Tom Knight) were capable of producing custom-ordered, programmable cells on demand. A synthetic yeast was even used to make beer in 2020, which tasted quite good.

This quest to engineer life has led to an appreciation for the fundamental building blocks of life itself. For example—what is the minimum number of genes needed for a cell to survive? Work from the J. Craig Venter Institute (JCVI) has begun to answer these questions by examining the simplest of organisms. Researchers (including Craig Venter) have been slowly removing genes from *Mycoplasm mycoides* (JCV-syn1.0), in an effort to get to the bare genetic minimum code set. Notably, *M. mycoides* is an organism that has very low gene content (~600 genes) and has shown that life can have a lean genetic form.

By 2016, in a seminal paper (by Hutchison et al.), the JCVI team had shown that *M. mycoides* only needed 473 genes to survive on minimal medium. While it is not clear what we might find on another planet, it seems this is close to the minimum bar required to be an independent, functioning, living organism on Earth. However, this is based on the current genetic code. Ongoing efforts by members of the synthetic biology community, as well as those in GP-write, have pioneered ideas for more types of genes, genomes, and even the fundamental building blocks of life's chemistry. All of these synthesized possibilities on Earth could actually resemble what we might find naturally occurring on other planets once we being to explore.

RECODING THE GENOME AND NEW GENETIC CODES

Earth's biology is primarily based on just four nucleotides, which are then transcribed into ribonucleotides (RNA) to move information around a cell. These substrates of life and the orientations they have fallen into over time to create the diverse life across Earth is really no different from a spilt bowl of alphabetical soup. Crashing to the ground, each noodle falls randomly next to another, forming words and elements on the floor; then a toddler (selection pressure) comes crawling along and selects the letters that make sense and that should be passed on (fitness). While the combinatorics of this fundamental framework can be vast, with the potential to create more distinct organisms than the sun will allot time for, the read-out of the current genetic code in humans is actually quite limited.

The "accidental evolution" that has formed life thus far is also the reason for its current constraints. These constraints are bound by the four core letters (nucleotides) that make up the alphabet of Earth's life, as well as its very grammar, constraining these four bases into sixty-four three-letter codons ($4^3 = 64$ codon combinations). If codons were four bases long, there would be 4^4 (256) combinations of codon-tRNAs. But, even though 256 is clearly a lot more than 64, is this increase actually better for life? This drastic increase would necessitate more tRNAs, which would mean fewer chances of identifying the correct match, would overall be somewhat wasteful for cellular resources, and would also require all tRNAs to be adjusted for appropriate binding. Current constraints and combinations actually come with varying levels of redundancy. These redundancies include three codons that all mean "stop" (essentially a period at the end of a sentence), as well as redundancy in AA tRNA matching (sixty codons for matching twenty amino acids).

This redundancy, found across all levels of Earth's life, can actually be leveraged in a number of distinct ways through "recoding" genomes. For example, since viruses' genomes include the redundant codons as part of their programming, a virus will be less effective (or even not at all effective) if a cell is recoded to only have one codon for a single amino acid. In 2016, George Church began the effort to create "virus-proof" cells through this recoding strategy. The same goal is now being implemented for GP-Write to make human cells that cannot be infected with viruses, which could then be used for growing pharmaceuticals and bioproducts for mass use. The narrowing of the genetic code can focus the efforts of an organism and give it the benefit of being immune to infection.

Whereas eliminating redundant codons in select organisms essentially creates a "secret code" that foreign invaders aren't privy to, generating a broader alphabet would lead to far more words in life's lexicon, which no non-engineered organism would possess. Instead of the four bases of the genetic code, couldn't we have more options beyond the natural base pairs? Yes.

Studying unnatural base pairs (UBPs) started in 1989, with Steven Benner, who made modified forms of cytosine and guanine, incorporated them into DNA molecules in vitro, and then got them to replicate reliably. In 2002, a Japanese research team led by Ichiro Hirao created a UBPs, with a purine (A,G) and pyridine (C,T) combination capable of transcription, and further incorporated these coded nonstandard amino acids (NAAs) into proteins. Their full names were 7-(2-thienyl) imidazo[4,5-b]pyridine (Ds) and pyrrole-2-carbaldehyde (Pa); for obvious reasons, people just refer to them as "Ds-Pa."

The long-term use of any new UBP and NAA system would need to be stable (not revert back to the natural bases) and capable of high precision replication to retain the code across generations. Some of these goals were met in 2006, with a six-letter framework called the artificially expanded genetic information system (AEGIS), also from Benner. It included the four basic "vanilla" nucleotides (A, C, G, T) alongside two additional nonstandard "flavored" nucleotides (Z and P). This new kind of ACGTZP DNA sequence base greatly increased the potential combinations of amino acids using the standard three bases per codon, 216 (6^3) versus our 64 (4^3).

In 2012 and 2014, Floyd Romesberg and his team at the Scripps Research institute packaged up this idea for delivery into a host cell. They made a plasmid containing natural base pairs along with a new UBP (called d5SICS-dNaM). This plasmid was integrated into the *E. coli* cells and propagated for several generations—the first example of a living organism passing along an expanded genetic code to subsequent generations. This success did not come easily, though, since over 300 UBP nonfunctional variants were tested until the cells could reliably replicate the new functional bases (d5SICS-dNaM) that nature had never before seen. The team also needed to add an "alien gene," specifically a gene that expressed nucleotide-triphosphate transporters (NTTs), the enzymes that efficiently import the triphosphates of both d5SICSTP and dNaMTP.

In 2019, Brenner, Shuichi Hoshika, and Nicole Leal went even further and produced an eight-letter genetic code, expanding its potential to 512 combinations (8^3), which was properly named "hachimoji DNA" (Japanese for "eight"). These four new bases (P, B, Z, and S) were derived from the same kind of nitrogenous structures as purines and pyrimidines. These structures enabled the nucleotides to be true UBPs and form hydrogen bonds to their pairs, specifically S-B and P-Z. However, there is more to a functional genome than its hydrogen pairs. As such, some challenges arose. These new additions to the genetic code changed the charge, structure, and stability of the double helix, which also augments the opening, closing, and control of the DNA and its resulting expression. Future work would be needed to engineer and optimize enzymes to control this new code.

These new forms of DNA not only show what kinds of life could exist on Earth through human engineering, but also elucidate how different life might look on other planets. On Earth, these alterations can be used to create improved and infection-resistant cells for the creation of novel biopolymers and therapeutics. These alterations will likely be incorporated into humans in a step-wise fashion, like other genetic engineering ideas. The first, and potentially near-term, application of recoded cells into humans would be through hematopoietic therapies. As an example, HSCs or precursor T-cells could be recoded and further engineered for their therapeutic intent (be it to fight cancer or to improve hemoglobin formation), thereby decreasing the chance of contamination in the engineered product and further preventing against infections in the patient post-transfusion. This would address one of the current largest risks of transfusions (infection), enable recoding in a quality-controlled environment, and could be oblated through the usage of safety systems (essentially self-destruct switches added into engineered cells). Once this is established and safe, entire germlines could potentially be reengineered and grown within the exowombs.

Our current work on recoding Earth's genome spans only a few decades. Given several hundred years, it is possible that entirely new adaptable systems could be deployed for other planets to combat their unique challenges. The constraints of Earth's life-forms, and how much we are capable of pushing their limits—be it through identifying the

minimum number of genes possible for an organism to thrive, or pushing the complexity of the usable alphabet to produce more complex
lexicons while still being able to quickly reproduce—give us an understanding of the *minimum* difference we can expect to see on other planets. Life on these other planets may be drastically different from here
on Earth given that it originated, grew, and changed under a different
set of planetary rules, or it could be based on the exact same fundamental rules as life on Earth, consisting of just four nucleotide and 64
codons, resulting in a universal law for how life emerges. Until we find
other life-forms that grow under the protection of a different celestial
body, we will not know, but time will tell.

Purposefully recoding the genomes of select organisms could open
up a new form of "nucleic-acid planetary protection." Within this paradigm, we could engineer organisms such that they could only survive
on the specific planet to which they are going, therefore preventing
accidental contamination from one planet to another. Further, after
thousands or millions of years, we could learn which nucleic-acid configuration and systems would be the most stable or adaptable within
these environments. These configurations can then be used to empiri-
cally define recoding strategies with the best chance to enable organism survival when sent on missions to planets with more unknown
properties far outside of our own solar system. Therefore, these systems
could be used in some organisms to protect accidental contamination
across worlds, while employed in different ways for other organisms to
maximize survival across a large variety of different worlds.

CROSS-SPECIES ORGANS

But life is more than just surviving; it's about thriving. We already know
from animal models and the Mason lab's work, led by Craig Westover,
that genetic elements identified in one organism can be functionally
incorporated into another. Further, we've explored cases of creating new
genes through the combinations of others to enable new cellular functions. But what about larger organ or tissue system-level changes? Could
we take an organ from one organism and incorporate it into another?

Would this enable the engineered organisms to do more than just survive, but to thrive? What would be required to make this a reality?

We don't need to look into a potential future for this application; we need only look back to 2017. To solve the human organ shortage, scientists at a company called eGenesis used CRISPR to remove twenty-five dangerous viruses from their pigs' genomes. When we try to use pig organs for transplants, common porcine endogenous retroviruses (PERVs) within the pig genome can lead to transplant rejection. Further integration of PERVs into the human genome can lead to immunodeficiency and even cancer, so either removing or inactivating the PERVs is required to make these therapies safe and scalable. However, this is not the only barrier to using other species organs as human transplants. Zoonotic (animal-to-animal) transmission risks of pathogens include mad cow disease, Ebola, and some coronaviruses (SARS-CoV-2), and is possibly even the source of HIV. Further, even if you remove all potential viruses from the donor animal, additional complications have been observed, including improper blood clotting and immune rejection.

In 2020, the CSO of eGenesis (Luhan Yang) collaborated with Qihan Bio (a Chinese-based, similar xenotransplantation company) to make pigs with forty-two specific genetic changes that could help with transplant rejection. She called them Pig 3.0, because this was their third version of an edited pig. They edited or inactivated 30 PERVs, plus three pig genes that could cause human immune rejection (GGTA1, CMAH, and B4GALNT2), and then added nine human genes that should make pig organs more compatible in a human host (hCD46, hCD55, hCD59, hB2M, hHLA-E, hCD47, hTHBD, hTFPI, and hCD39).

Upon further examination, the Pig 3.0 animals seemed just fine. Not only were they observed to be perfectly fertile, but they also produced a normal litter size that contained the genetic modification in a normal Mendelian fashion. The litters were capable of evading human antibodies (IgG and IgM) with a 90 percent reduction in binding. Moreover, the 3.0 pigs showed higher resistance to NK-based cell killing, slight suppression of phagocytosis by macrophages (10 percent), and improved blood-clotting. Overall, their data showed that their edits performed as expected, even across species. Such engineered designs are expected to work even better with decades or hundreds of years more of development, enabling universal animal-to-animal transplantation.

But how will this new world order look if one species is edited to serve as a host for another species' body parts? Such a world raises the ethical question of whether this process violates the rights of that animal. According to the philosopher Peter Singer, treating one species as morally more important than another (speciesism) is unacceptable. It is a violation of the organism's rights, and it represents bigotry across species, just like racism is an irrational and unethical bigotry within one species. Basically, we are not affording the pigs a chance to have their own agency and avoid this possibly painful process.

But what if the organ development occurred without any suffering for the pig, surgery is painless, and the process enabled a global decrease in the total amount of human (or even other animal) suffering? If there is no suffering, and deontogenic, Kantian, and utilitarian ethics are upheld, then it might be justified.

Yet some may argue that the act of engineering a species for another species' betterment is, in itself, wrong—no matter what the situation. They might argue that it is exploitative and/or that we should leave Nature alone. However, a simple thought experiment can help clarify the question. Imagine there is a car stuck on a train track with a stranger inside of it. To stop the train and prevent an accident, you only need to flip a switch, thereby saving the stranger's life and everyone on board.

Should you flip the switch? Most would say yes. In that case, your intervention changes the fate of certain death, much like engineering an organism away from certain disease. But what if you are forced to jump into an oily pool and ruin a $5,000 suit to flip the same switch? Remarkably, even though the only additional negative here is strictly financial, more people would now say no. However, the vast majority would say yes because they value a human life more than $5,000. Your moral duty requires you to act, even if you did not create the situation. The question then becomes: What is forfeited for this outcome? What is the baseline comparison required to come to a conclusion on whether or not to engineer an animal in the above scenario?

The engineered, organ-producing animal would either be created in such a way that the process of removing the organ will be a simple and painless surgery that enables the animal to live life as well as before (saving one more life), or a painless surgery that would result in the sacrifice of the animal (likely saving multiple other lives). In both of these scenarios, we have acted to cause a neutral or net gain in life. However, what about the second scenario where the animal is sacrificed? This baseline comparison should, in fact, not be the absence of engineering but instead the absence of human intervention. Animals are constantly being eaten alive by other animals, clawing at the ground for their freedom, gasping for their last breath.

In this case, the engineered animal would likely have a better life, more "humane" death (either by living out its post-surgery life protected from predators or asleep during surgery, roaming open fields), and further enable the prolonged survival of other species. Given that this increased life span may enable the prolonged survival of the meta-species, deontogenic ethics would dictate that this action is ethical.

ETHICAL ERADICATION

In fact, this might not go far enough. All around us are animals that suffer from accidents of evolution (back when it was unguided). Usually, we don't notice or even know when animals around the world are suffering; we don't even know their state of mind, as is well exemplified by Thomas Nagel's work in the essay "What Is It like to Be a Bat?" But we can know some aspects of their pain, which has been well described by Kevin Esvelt from MIT, a leader in genetic engineering, gene drives, and ethical uses of genetic engineering (at the Sculpting Evolution Laboratory).

Some of Esvelt's work has pioneered a single, disruptive question: when are we *ethically bound* to genetically edit creatures in the world around us? Some examples are easy to support, like the eradication of smallpox. This has clearly been a benefit for humanity, even if not so much for the bacteria. We do have the genome of smallpox still banked, so if for some reason we ever need to, we could de-extinct it.

But other species are more complicated. For example, there is a New World screwworm fly (*Cochliomyia hominivorax*) that buzzes around most livestock farms in the Americas. The female flies often lay their eggs in the open wounds or skin of the livestock, leaving behind maggots that devour healthy and injured animal tissue alike, digging into the flesh of their hosts until they fall down from the pain. But before the animals fall down, the maggots in the wounds emit pheromones that serve as a signal to new females, to bring more flies and maggots. When this fly infestation happens in humans, it is so painful that patients require morphine before the doctor can even *gently touch* the wound.

Sadly, animals are not as lucky as humans to get treatment or pain relief. In any given minute, millions of animals around the world, especially in the Americas, are being eaten alive by these screwworm flies and their maggots writhing around in soft tissue. It was only recently that this was fought and stopped with some sterile insect techniques (e.g., releasing sterile males into the environment). Notably, this was not done from some high-level, altruistic drive to keep the animals or people safe, but rather for the cost: the flies are estimated to cost $4 billion in lost revenue for the farms, across cattle, sheep, and goats.

Yet the sterile insect technique is not easy to deploy, especially in mountainous areas or rough terrain. What is needed is a self-propagating genetic element that could make its way through the entire population and change the species . . . a "gene drive." Gene drives, as described by Esvelt, Church, and others in the field, are a natural phenomenon that occurs when a genetic element reliably spreads through a population, even if the fitness and survival of the recipient organism is slightly reduced. Once CRISPR genome editing emerged, it became possible to replace the original sequence with the edited version *and* an encoded copy of the CRISPR system and have this genetic element propagate through the ecosystem. Thus, with a gene drive, it is possible to remove the cause of suffering for hundreds of millions of livestock, small animals, and people, but it would likely change the ecosystem in ways we cannot yet predict. As always, there is a risk we will create more problems than we solve. Regardless, ameliorating this widespread suffering, even if it requires the eradication of the source and addition of another non-harmful organism to stabilize the ecosystem, is the most ethical thing to do.

In both gene drives within insects and editing astronauts to keep them safe, we have the ability to intervene to avoid suffering and ensure safety. By 2200, it is likely that this method will be tried extensively in laboratories and that new types of gene drives will be created and also tested, including the ability to safeguard them. Instead of just physical or ecological confinement, "molecular safeguarding" can also be pioneered with two methods well-tested by Jackson Champer and Philipp Messer at Cornell University. The first idea is to use a "synthetic target site" drive, which creates engineered loci in the DNA of the target species that do not occur in the wild. The second method is a "split drive," where the drive alone cannot work, since it lacks its own endonuclease and thus needs one from another source.

Other control mechanisms could be deployed from 2050 to 2200 that should also reduce the risk. These include "sensitization drives" that only work in the presence of a particular chemical, purposely "unstable drives" that limit how many drive-copying events can occur, "interacting drives" that are suppressed when they encounter another gene signature (e.g., another gene drive), and an "immunizing drive" that could protect a subpopulation, such as males or a genetic subset. While it is not clear which methods will be the most efficacious, the agency of being able to control our own evolution, and that of the environment, will likely develop very quickly.

Based on the previously discussed scenarios and deontogenic ethics, there are at least three examples where we may be ethically *required* to edit genomes:

1. Reproduction of life: couples who want to have children but cannot (including for same-sex couples or genetic diseases);
2. Survival of life: humans being sent to hostile worlds (protected astronauts);
3. Quality of life: preventable animal suffering (e.g., the New World screwworm).

In these cases, we are ethically obligated at least to offer the genetic engineering.

ENHANCED PERCEPTION

As we approach the year 2201, human enhancements will be wide-spread, with more novel ones beginning to emerge. These could include fun abilities, such as the reintroduction of tails, to more potentially useful enhancements, such as improving how our vision works so that we can perceive more spectrums of light. While these ideas were found only in the realm of science fiction in the early 2000s, they could be prominent in real life by 2201.

For example, the means by which we use the rods and cones in our eyes could be radically improved. When you look at light, rod and cone cells in your eyes absorb photons, convert these ionic fluxes into signals, and then launch a long multigene and multicell cascade that results in the colors and shapes you see. One of the pivotal steps in the rod cells requires the use of cyclic guanosine monophosphate (cGMP) to carry information through the cytoplasm and to freely floating retinal discs—acting like a molecular helicopter landing on a moving helipad. The plasma membrane is continuous in cones and discs, but they are still using cGMP. This results in a large waste of energy, like taking your helicopter to visit your next-door neighbor.

Another crucial component of human vision is a molecule called retinal, which is derived from vitamin A, that helps propagate the perception of light after the interaction of a photon with a specific wavelength. Without vitamin A, circuits within human eyes begin to degrade and lose vision at night, eventually leading to blindness. Yet human cells, including both rods and cones, do not synthesize their own vitamin A. Oddly, this crucial component to life is entirely derived from diet (this issue is solved in the next chapter).

The overall capture of light could further be improved through studying animals which forage in low-light environments. Some animals have a mirror-like tissue layer in the back of their eyes, called the *tapetum lucidum*, that reflects light back through the retina, increasing the amount of light that it can capture and use to make an image. This tissue layer is already found in cats, dogs, some deep-sea animals, and some primates, such as aye-ayes. In principle, humans could also get a *tapetum lucidum*, but this would be at the cost of losing some focus.

NEW EYES FOR NEW PLANETS

To solve the focus issue, some nocturnal mammals have altered the overall structure of their eye cells, which both enables their lifestyle and could serve as a guide for future humans. For example, the nuclear pattern within the rod cells could be flipped, with the heterochromatin in the center of the nuclei and euchromatin and other transcription factors along the border of the cell. The thickness of this layer could also be increased to "stack the deck" with rod cells. This stacked deck would therefore increase the odds of light hitting necessary photoreceptors resulting in enhanced light capture with no loss of focus. Engineering eyes to be larger, with this layout, would enhance light capture even further. We could even add more segments to eyes (compound eyes, as in insects), instead of being limited to just one.

As alluded to above, our processing of light is highly specific to its wavelength. Cones could be further engineered to respond to additional wavelengths of light, though the interpretation of these newly absorbed photons and perception would require advanced multicellular engineering of the overall electric circuitry of the eye and brain. Solving these electrical limitations would enable an entirely new perception of the world and universe (e.g., thermographic or infrared [IR] vision).

Both the potential and challenges of enhancing human sight can be elucidated through studying the numerous other animals that see differently than we do. Several cold-blooded animals can perceive infrared light, including snakes, which have heat sensors along their upper and lower jaws. Mosquitoes can (unfortunately for us) also see the CO_2 and heat from the mammals on which they feed, enabling highly precise blood sucking. Some fish (goldfish, salmon, piranha, and cichlid) can activate infrared vision to help navigate murky waters. Even frogs have infrared vision, which they activate through the use of vitamin A and an enzyme called Cyp27c1.

Another option would be to create IR or near-infrared (NIR) vision so we could see heat signatures in the world around us, like within mosquitoes. The human eye has evolved to specifically detect light between the visible wavelengths (400–700 nm), but work from Gang

Han at the University of Massachusetts Medical School has shown it might be possible to extend our vision to 750–1400 nm (near infrared). This would work like thermal-imaging cameras, which detect IR radiation given off by objects. In 2019, Han and colleagues injected a nanomaterial called upconversion nanoparticles (UCNPs) behind the retinas of mice. These nanoparticles held several rare-earth elements (erbium and ytterbium) capable of converting low-energy photons from NIR light into higher-energy green light, which the mice then interpreted as a normal visible light. This was made possible by attaching a protein that bound to glucose on the photoreceptors, which then changed the substrate. The engineered mice were capable of navigating a Y-shaped, NIR-lit tank, whereas the non-engineered mice, blind to the NIR signal, could not find their way.

Perhaps most importantly, the UCNPs seemed safe. They functioned in the mice's eyes for at least ten weeks and did not cause any noticeable side effects. Thermographic vision in a mammal could, in principle, be engineered in a similar fashion with organic dyes, be adjusted to emit more colors, and, theoretically, be added early in life. This could be critical for long-duration missions that would take an astronaut to a planet far away from the sun (e.g., Titan), where low-light and IR/NIR vision would be immensely useful.

Even the wiring of the brain itself could be improved. As an example, when comparing mammals across the evolutionary tree (phylogenetically), distinct differences emerge in the "higher mammals" within the sensory and motor wiring in their cerebral cortex (which controls basic movement), as well as a distinct corticospinal tract design (the motor neuron roadway which descends from the brain). Specifically, the number of corticospinal axon terminals, which reach out to spinal motor neurons, continues to increase from prosimians to apes to humans, and may even influence how/when language develops (as shown by Erich Jarvis). Hence, further increasing this number may allow for the development of more refined motor skills, which could improve survival and adaptability on other planets. We can expand the limits of what life can detect, and then begin to expand the limits of life itself.

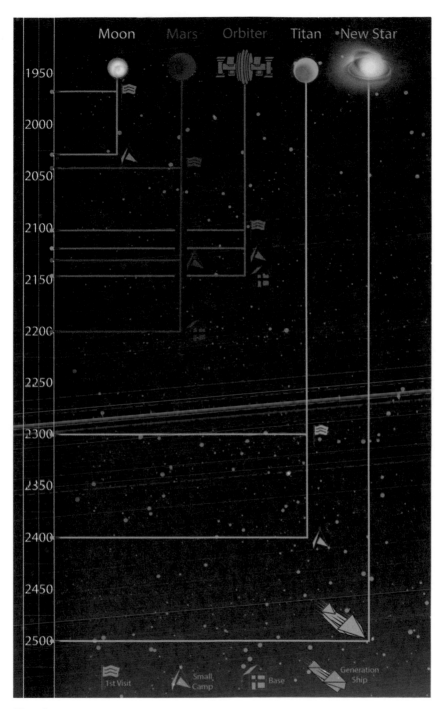

Plate 1

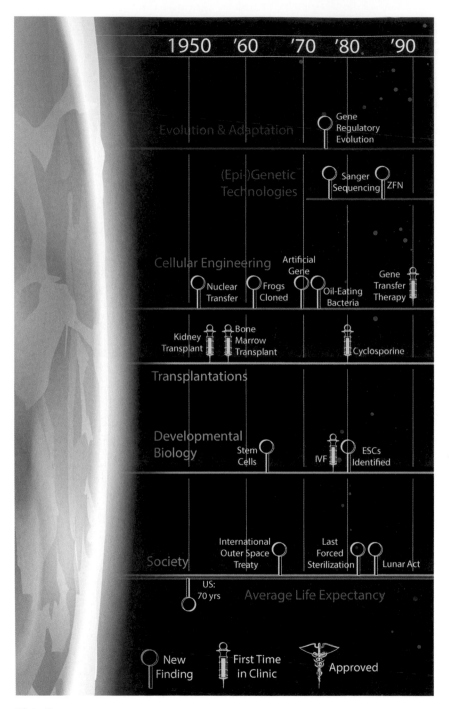

Plate 2

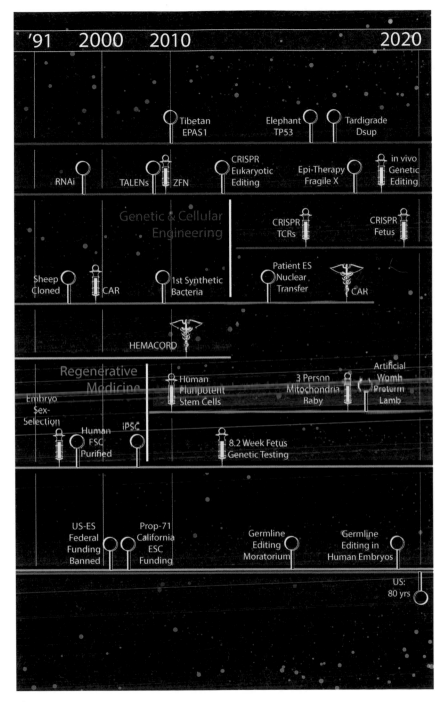

Plate 3

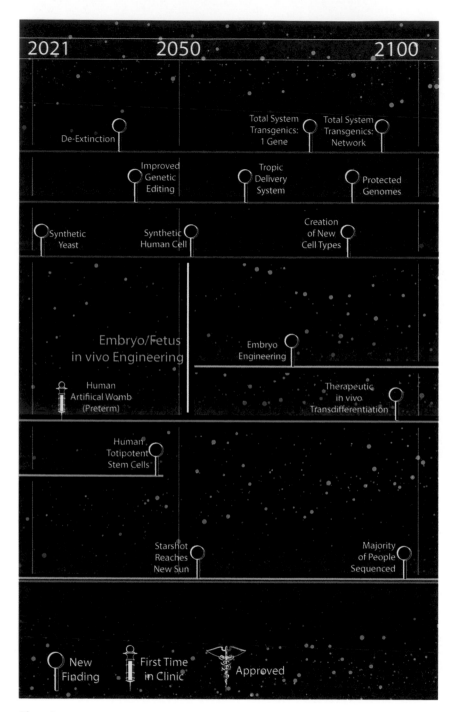

Plate 4

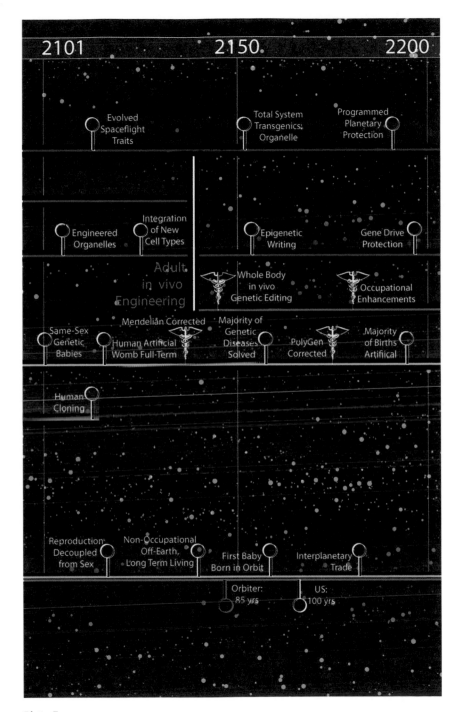

Plate 5

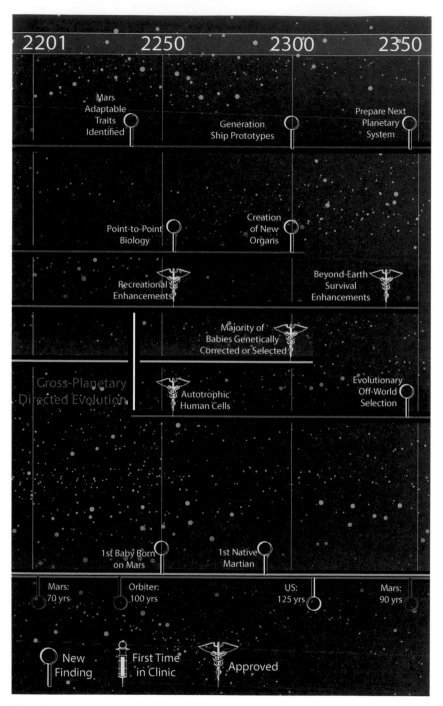

Plate 6

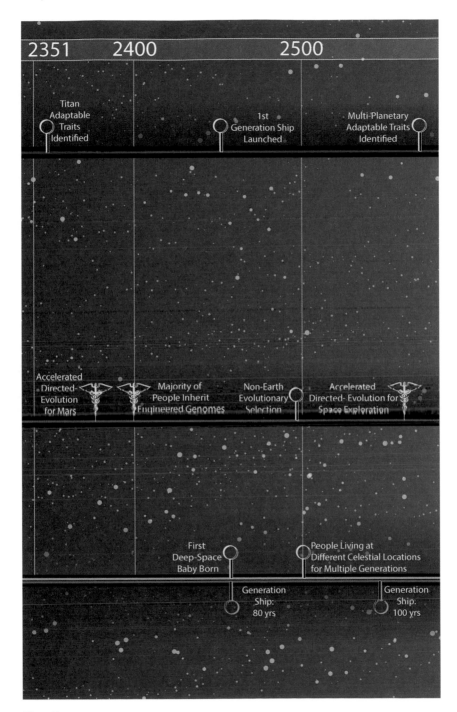

2351   2400   2500

Titan
Adaptable
Traits
Identified

1st
Generation Ship
Launched

Multi-Planetary
Adaptable Traits
Identified

Accelerated
Directed-
Evolution
for Mars

Majority of
People Inherit
Engineered Genomes

Non-Earth
Evolutionary
Selection

Accelerated
Directed- Evolution for
Space Exploration

First
Deep-Space
Baby Born

People Living at
Different Celestial Locations
for Multiple Generations

Generation
Ship:
80 yrs

Generation
Ship:
100 yrs

Plate 7

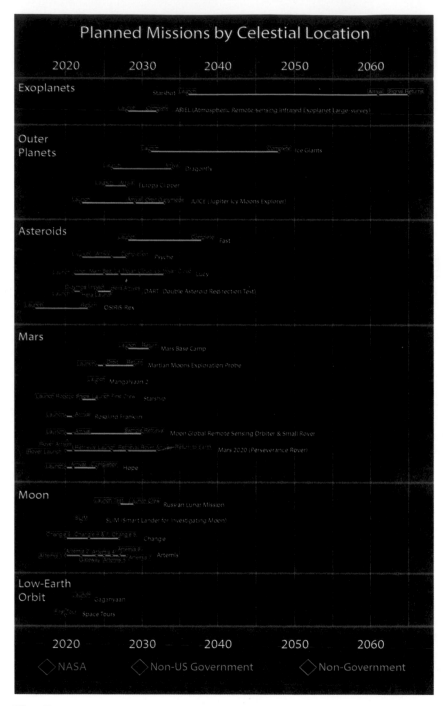

Plate 8

Near-term missions for space exploration and human settlement: Missions are grouped by locations, including low-earth orbit (LEO), the Moon, Mars, asteroids, outer planets within our solar system, and exoplanets. Years are given on the bottom and mission details are highlighted on each line.

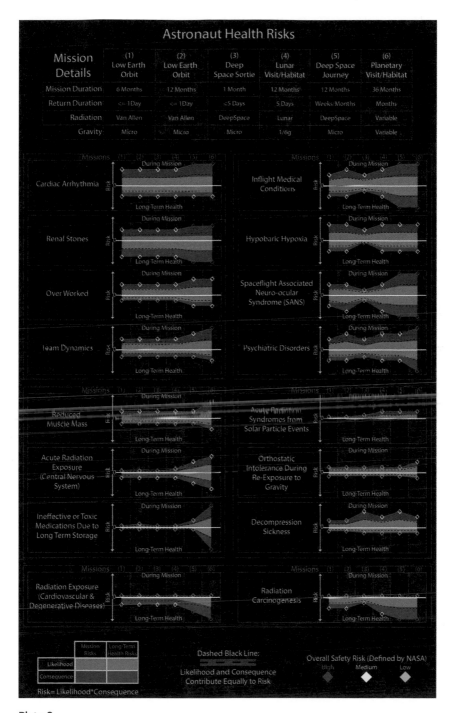

Plate 9

Risk factors for astronauts: Current risks are stratified by high (red), medium with and without mitigation (yellow), and low (green), based on mission details such as where the astronauts will be and for how long. X-axes represent defined missions (1–6) where Y-axis shows relative risk for the mission (purple) or long-term health (blue), based on likelihood (lighter) and consequence (darker).

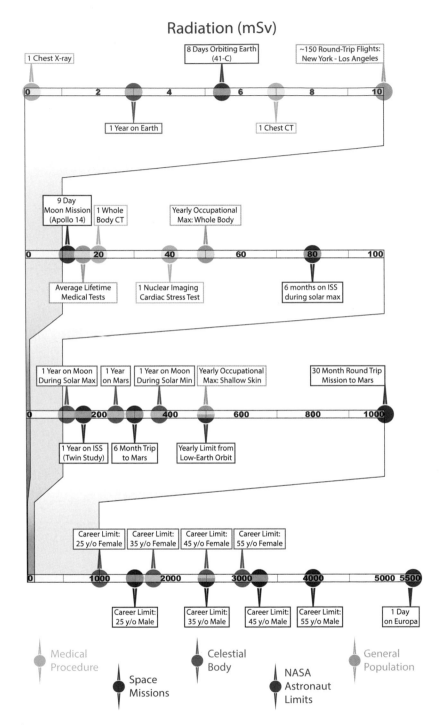

Radiation (mSv)

Plate 10

Radiation risks and levels: Estimated levels of radiation for varying-duration missions and exposures, including medical procedures (green), the impact on various celestial bodies (blue), specific space missions (purple), NASA-recommended astronaut limits (red), and the general recommendations for terrestrial radiation (yellow).

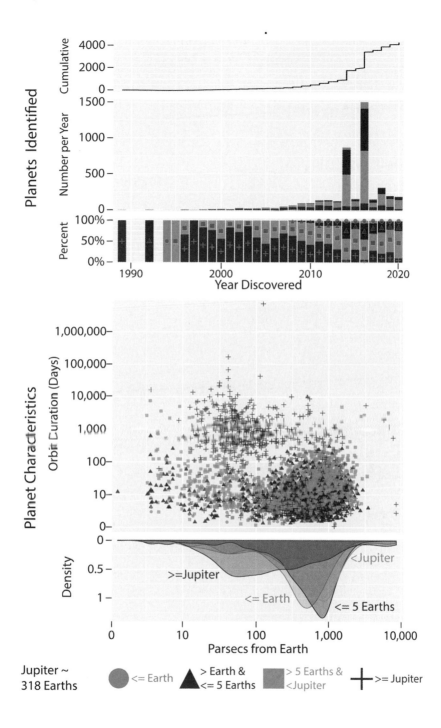

Planets Identified

Cumulative

Number per Year

Percent

Year Discovered

Planet Characteristics

Orbit Duration (Days)

Density

Parsecs from Earth

Jupiter ~
318 Earths

<= Earth

> Earth &
<= 5 Earths

> 5 Earths &
<Jupiter

>= Jupiter

<Jupiter

>=Jupiter

<= Earth

<= 5 Earths

Plate 11

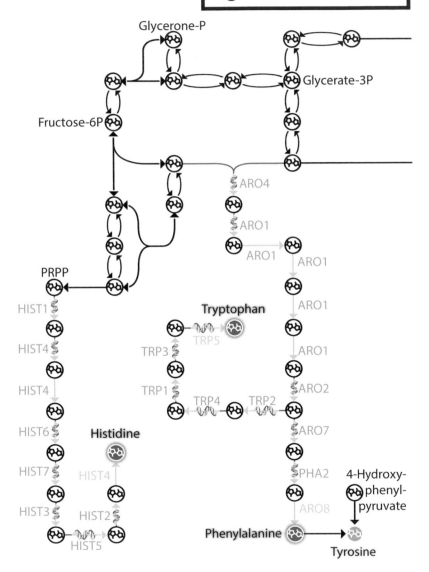

Fructose-Derived "Essential" Amino Acids

Molecules

- Current Essential Amino Acid
- Non-Essential Amino Acid
- Intermediate

Glycerone-P

Glycerate-3P

Fructose-6P

PRPP

HIST1
HIST4
HIST4
HIST6
HIST7
HIST3

Histidine
HIST4
HIST2
HIST5

Tryptophan
TRP5
TRP3
TRP1
TRP4 TRP2

ARO4
ARO1
ARO1
ARO1
ARO1
ARO2
ARO7
PHA2
ARO8

4-Hydroxy-phenyl-pyruvate

Phenylalanine

Tyrosine

Plates 12, 13

Pathways and steps for amino-acid biosynthesis in humans: Human biosynthesis pathway reimagined to enable the production of all amino acids with as few steps as possible. Black lines indicate steps already present in humans, blue and red lines indicate genes derived from other species. Green circles represent "non-essential" amino acids,

Acid Biosynthesis Pathway

Serine-Derived "Essential" Amino Acids

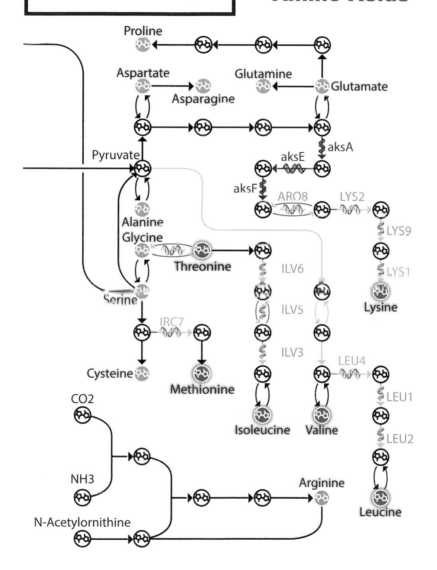

which humans are currently capable of synthesizing, where purple "current essential" amino acids represent the amino acids that need further engineering to be produced by humans. The estimated number of biochemical steps and their associated genes are shown along with the input needed for synthesis.

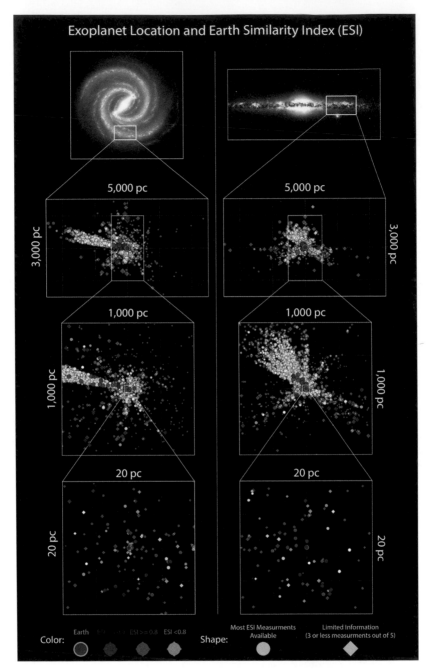

Plate 14

The location and similarity of all identified exoplanets: Most putative exoplanets that could be used for settlement by generation ships are within dozens or hundreds of parsecs (3.26 light-years) from Earth, either when examined from the top of the Milky Way (left) or the side view of the Milky Way (right). The ESI for a variety of planets can be calculated based on the metrics of equilibrium temperature, density, solar flux, radius, and escape velocity. The highest-quality candidates are ≥0.9 (blue), secondary candidates are <0.9 but ≥0.8 (purple), and the lower-quality candidates are <0.8 (gray). Exoplanets with only 3 or less metrics for ESI calculation are diamonds; those with 4 or all 5 metrics are circles. Data from https://exoplanetarchive.ipac.caltech.edu/.

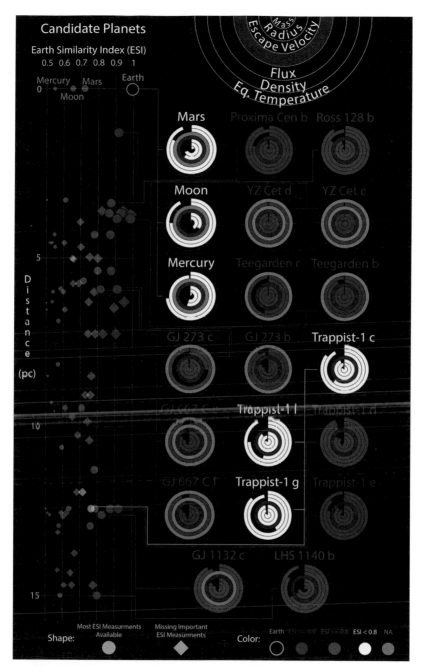

Plate 15

Best extracellular worlds based on location and similarity: ESI for candidate planets is shown relative along with their distance to Earth (y-axis). The ESI for a variety of planets can be calculated based on the metrics of equilibrium temperature, density, solar flux, radius, and escape velocity. The highest-quality candidates are ≥0.9 (blue), secondary candidates are <0.9 but ≥0.8 (purple), and the lower-quality candidates are <0.8 (white). Relative values are displayed as filled-in rows within the circular plots; light gray references missing data.

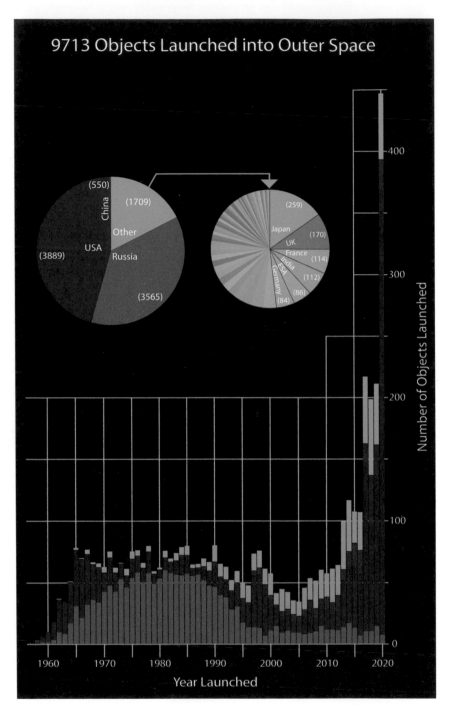

Plate 16

The number of objects launched into space, summarized by the year of launch and country. This includes Russia (purple), the United States (blue), China (red), and other (green) countries or agencies, which include Japan, France, India, the United Kingdom (UK), European Space Agency (ESA), Germany, Intelsat, and others.

8

PHASE 6: EXPANDING THE LIMITS OF LIFE (2201–2250)

The best way to predict the future is to invent it.

—Alan Kay

In 2201, many different enhancements will be created to improve life on Earth, as well as an astronaut's ability to complete a specific mission or thrive on a new planet. The enhancements that will be created and how they will be incorporated will depend on where people are going. The physical and environmental challenges of the worlds they visit or make their new homes will dictate the requisite genetic and molecular engineering. Some homes will be easy, but some will be difficult, depending on how far the reach of life can go.

THE LIMITS OF LIFE

Almost all of what is expected for life's features in this solar system is built around what we know about the range of life's possible adaptations on Earth. But this raises key questions: What do we know about the limits of life? How far can it go, and how does this translate into the ability to adapt and thrive?

To address this, we again look to our favorite friends: extremophiles. Some extremophiles have adapted to their environments to such a degree that they actually require these environments to thrive. Others, however, are capable of tolerating these extreme environments but would prefer to live in less intense places (extremotolerant), enabling them to visit their more *extreme* friends. Ideally, through further genetic and cellular engineering, we can make humans and astronauts more extremotolerant. Though there may be a perfect replica of Earth for us to visit one day, either naturally occurring or through our own innovation, living on other celestial bodies in our solar system will require some extensive engineering. Whereas some areas on Mars may fit within Earth's temperature extremes, other planets such as Venus are far hotter than even the hottest recorded temperature on Earth (figure 8.1). If we are to enable these extremotolerant abilities within humans, we must first examine Earth's extremophiles.

The Extreme Microbiome Project, led by Scott Tighe at the University of Vermont, Kasthuri Venkateswaran at NASA, and the Mason lab at Weill Cornell Medicine, is a group of scientists hard at work detailing Earth's extremophiles. This work not only shows us the conditions under which Earth's life can survive, but also reveals the biochemical mechanisms that enable this life to be possible. Once these biological functions of adaptation have been understood, we can then begin to translate them into other systems, such as human bodies, or even other

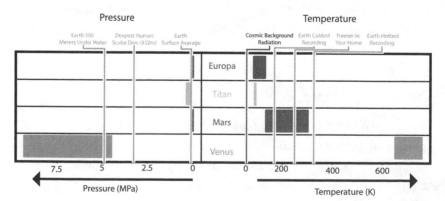

8.1 Survival gradients: Comparing pressure and temperature, it is clear that Earth is in a special place in our solar system.

organisms that we will undoubtedly send to other worlds to ensure their survival.

Extremophiles live across Earth in many different environments, with many types of stressors, including: high/low temperature (thermophile/psychrophile), high pressure (barophile or piezophile), high salinity (halophile), high/low pH (alkaliphile/acidophile), and high radiation (radiophile). Further, endoliths can persist in the microscopic spaces deep within Earth's rocks, aquifers, and fissures, which are also the most likely spots holding life on Mars and other celestial bodies. However, it is also possible that life on Earth began with thermophiles along the hydrothermal vents in the ocean floor (black smokers), and now can be been found in hot springs, hydrothermal vents, and deep-sea vents (see below). Notably, *Pyrococcus fumaris* is a thermophile that was found replicating at 113°C near the walls of black smokers. Other black smokers in China are shown to have *arqueobacteria*, which can survive at 400°C using chemosynthesis to produce H_2S.

Psychrophiles have specific adaptations to the cold, making them love Arctic and Antarctic climates. Work from Pabulo Rampelotto has shown that their proteins, in general, have more glycine, maintain greater flexibility, reduce their intramolecular interactions with other proteins, and come in smaller fragments—all to avoid freezing. One paper in 2014 (by Christner et al.) found some archaeal species in Antarctica that seem to require only NH_4^+ and CH_4 for survival, possibly persisting for millions of years in the total absence of sunlight or wind; these adaptations indicate they could even survive on Titan.

Radiophiles can resist high levels of radiation (both HZE and nuclear). The "superman of bacteria," *Deinococcus radiodurans*, can withstand radiation up to 5,000 gray (Gy), with no loss of viability, while *Thermococcus gammatolerans* can survive even higher levels (30,000 Gy). Both can be found hanging out in the cooling waters of nuclear-power plants, as if lounging on a beach chair in the radioactive, microbiological Bahamas. Just as superman has many powers, so do these bacteria—they can also survive extreme cold, desiccation, a complete vacuum, and even low levels of pH. They are thus known as "polyextremophiles," basically polygamous but for extreme states.

By 2201, a fairly complete catalog of Earth's adaptations and ability to thrive will exist. This will elucidate the genetic source of extremophile powers and be used as a framework to share these powers with other species. While these ideas are already being tested in 2021, such as our own lab's work embedding tardigrade genes into human cells to enable radiation resistance, this full catalog will enable the usage of more powers in more profound ways. These abilities can further be integrated into human cells on artificial minichromosomes, which would enable long-term persistence, no alterations to the protected human genome, and even allow for easy removal. This type of an advanced genetic incorporation system would enable the usage of highly specific modifications at precise timing for unique situations. Missions to reach planets or moons farther away, including Titan, will have started by 2250 and include new ways to see, and make, light and energy.

ENGINEERED LIGHT OR NO LIGHT

The biological rhythm of our sleep and wake cycles—the circadian rhythm—may be better maintained if the artificial light that surrounds us re-creates that of the sun or moon. The capacity to engineer light down to the specific nanometer of needed frequencies will help plants grow, humans work, and microbial species be controlled. The farther we get from the planet on which we grew up, the harder light will be to come by.

Quantum dots (qDots) are small, nanometer-sized engineered crystals that can emit light of various colors, depending on their size. They are built from the advances that led the revolution in the semiconductor industry, and now enable a full range of colors to be seen and produced (e.g., at Nano-Lit Technologies with Sarah Morgan). These qDots are now found in solar cells, fluorescent biological labels, and even overhead lighting in hospitals and offices. The ISS already has some of them being piloted out in several modules, as a means to improve how people live and work on the station. If the plan is to have a short trip away from Earth, technologies which enable Earth-like ambiances, like the rising and setting sun, will allow for easy transitions and better quality of life. But will this normalization of a new planet's light to Earth's standard cycle always be necessary?

Fortunately, many organisms on Earth have adapted to the loss of light. If they can, perhaps we can, too. Creatures that live in the depths of the ocean are free from the constraints of the sun's light. One compelling example is the anglerfish, which uses bioluminescence to create a decoy light that hangs in front of its face to capture prey—a unique form of predation. It's as if a fishing rod and a Venus flytrap had a baby. In the dark depths on the bottom of the ocean, a faint light in the distance may seem enticing to some. As soon as the prey gets close enough, the anglerfish (also known as the lantern fish) will strike. The anglerfish is just one of dozens of known species that have survived for 100–130 million years (based on mitochondrial sequencing studies) at the bottom of a cold, dark ocean. Thus, entire ecosystems have managed to live in the complete absence of light and near-freezing water (273K), at 3–4 percent salinity, long before humans emerged in their current form (about 6 million years ago).

Saturn's moon Enceladus likely has an entire subsurface ocean, which is extremely cold (about 100K) everywhere except for where plumes of heated silica shoot out of its center. There, ocean temperatures can reach about 190°F (360K), which is close to the temperature of the hydrothermal vents in Earth's oceans. Research from Sean Hsu and colleagues in 2015 showed that the plumes of heated silica and saltwater may represent an exciting place for life to develop and even thrive.

This potential for life is based on adaptations found in Earth's hydrothermal vents. These vents are usually found near "planetary push-pull" areas, such as volcanically active areas on the ocean floor where tectonic plates are moving apart. Enceladus has the push-pull of Saturn's gravity and orbital dynamics to thank for its plumes of silica activity, whereas Earth's tectonic plates are still in motion and are pushing and pulling along the mantle atop a still-liquid iron core. In both cases, the continuous planetary squeezing helps life emerge, as if it is being massaged into existence.

Entire ecosystems have been growing and thriving for millions of years in terrestrial plumes, creating entire new sets of life and biochemistry processes. These ecosystems include organisms such as Siboglinid tube worms, which can be larger, more slimy, and (potentially) more cuddly than an average human. The vents can reach temperatures of 400°C (673K), and around them swim various forms of shrimp, clams, and (of course) microbes. These microbes have lived without the sun's light for possibly billions of years, using chemosynthesis to manipulate carbon into different forms. These microbes convert nutrients into organic matter through oxidation (loss of electrons) of inorganic (no carbon) compounds, such as hydrogen gas from the vents, hydrogen sulfide, or methane. They use these molecules as a source of energy and electron transfer to survive.

The microbes of the deep even have symbiotic "buddy systems" similar to animals on the surface of Earth. The giant tube worms (*Riftia pachyptila*) of the deep-sea vents obtain necessary chemicals from an unusual, and personal, place. Gammaproteobacteria live inside of, and are therefore protected by, these giant tube worms and create organic compounds, which enable the worms to live without a digestive system. Because sulfur is only present in the extremely hot vent fluids, and oxygen is only found in the adjacent, cold seawater, the bacteria are able to acquire needed chemicals to live, all while hitching a ride (and while filling an apparent niche in the world). These "thioautotrophs" are the means by which giant tube worms live, and the pair make a beautiful example of a symbiotic relationship, which can serve as a model for Titan or other worlds.

Yet this symbiosis is not an easy thing to achieve. Sulfur reacts spontaneously with oxygen to form oxides, just as iron reacts with oxygen to form rust on a shovel. As such, thioautotrophs are normally only found at the interface between the ocean and the atmosphere, an environment which helps aid in their fight against sulfur "rusting." In the deep ocean, this environmental interface can also be found inside these large worms, enabling the possibility of this symbiotic relationship with gammabacteria.

Based on the diverse life thriving in the extreme reaches of Earth, it would almost be surprising if there is not *some* kind of life on Enceladus or other planets with liquid water. Even if life has not formed in these locations on its own, some of Earth's extremophiles may be able to thrive if we were to gently, and purposefully, place them there.

PROTOTROPHIC HUMANS

In many ways, humans are sad, frail moochers compared to many extremophiles and bacteria capable of creating all of their own required metabolites. Through undirected, volitionless evolution, humans today are unable to synthesize nine of the twenty amino acids needed for their survival. Therefore, to survive, humans must acquire those nine essential amino acids (histidine, isoleucine, leucine, lysine, methionine, phenylalanine, threonine, tryptophan, and valine) from their diet.

Unfortunately, human molecular ineptitude does not end there. Several essential vitamins needed by the human body are not made by human cells, including vitamins A (as described above in chapter 7 for eyes), B_1 (thiamine), B_2 (riboflavin), B_5 (pantothenic acid), B_6 (pyridoxine), B_7 (biotin), B_9 (folate), B_{12} (cobalamin), E, and K. Moreover, even though certain vitamins can be produced by humans (e.g., vitamins B_3 [niacin] and D), they are often not generated at sufficient levels and require significant supplementation through food or the gut microbiome. Through evolutionary selection and drift, humans and most other complex metazoans have lost the capacity to synthesize many of their own essential amino acids and metabolites. So, couldn't we just update the human genome with the missing steps required to synthesize these essential molecules?

To this end, Harris Wang and Jef Boeke began to create these updated human cells in 2020, through adding components of vitamin synthesis into human cells. Though engineering the entire catalog will take quite a lot of effort, the roadmap is already laid out for us to follow from studying other, more self-reliant, species. Some molecules will likely be quite easy to reincorporate, such as valine, which only requires the addition of four steps. Others, however, are much more complicated, such as introducing a twenty-two-step pathway to produce vitamin A. Tryptophan is estimated to take sixteen steps, phenylalanine may take

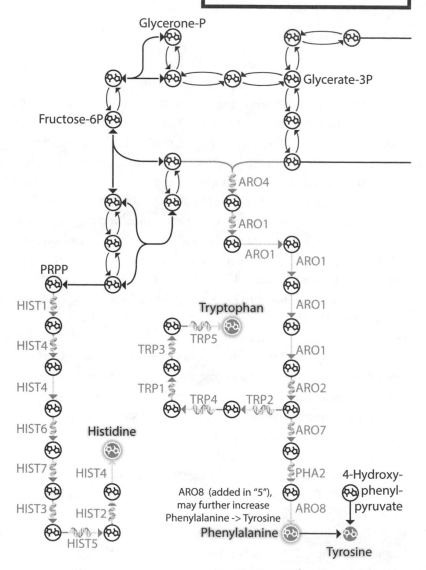

Fructose-Derived "Essential" Amino Acids

Molecules
- Current Essential Amino Acid
- Non-Essential Amino Acid
- Intermediate

Glycerone-P

Glycerate-3P

Fructose-6P

PRPP

ARO4

ARO1

ARO1

ARO1

HIST1

HIST4

HIST4

HIST6

HIST7

HIST3

Tryptophan

TRP5

TRP3

TRP1

TRP4   TRP2

Histidine

HIST4

HIST2

HIST5

ARO1

ARO1

ARO2

ARO7

PHA2

ARO8 (added in "5"), may further increase Phenylalanine -> Tyrosine

Phenylalanine

ARO8

4-Hydroxy-phenyl-pyruvate

Tyrosine

8.2 Pathways and steps for amino-acid biosynthesis in humans: Human biosynthesis pathway reimagined to enable the production of all amino-acids with as few steps as possible. Black lines indicate steps already present in humans; gray lines indicate genes derived from other species. Gray circles with no outer ring represent "non-essential"

Acid Biosynthesis Pathway

Serine-Derived "Essential" Amino Acids

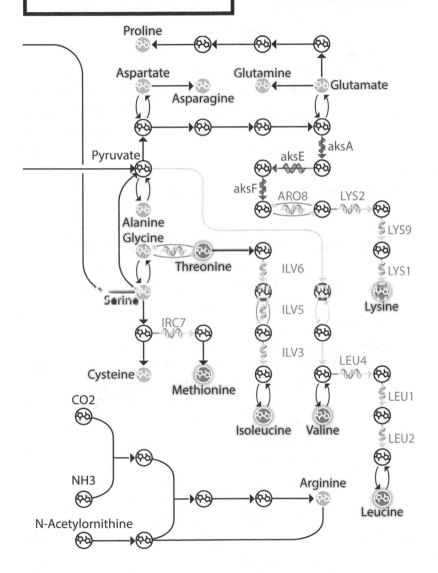

amino acids, which humans are currently capable of synthesizing, where "current essential" amino acids represent the amino acids that need further engineering to be produced by humans. The estimated number of biochemical steps and their associated genes are shown along with the input needed for synthesis. (See color plates 12 and 13.)

about thirteen steps, and even the relatively simple histidine requires ten serial steps of creation within a human cell. Altogether, for all these vitamins and amino acids, it may only be a tweak of 230 genes, which generates significant excitement in the open-source hardware, "do-it-yourself biohacking and practical transhumanism" wiki (http://diyhpl .us/diyhpluswiki/).

Long-forgotten genetic relics within our own genome could even be reactivated to enable this molecular self-reliance, instead of requiring the placement of entirely new genes. A dead pseudogene for vitamin C lies silently in our genomes, as well as that of other *Haplorhini* ("dry-nosed") primates, just waiting to be resurrected. However, *Strepsirrhini* ("moist-nosed") primates such as lemurs still have an active, functional copy. Scurvy, once a challenge for fifteenth- and sixteenth-century explorers on the oceans due to the lack of vitamin C, would also be a challenge in the far reaches of space, depending on available nutrients. However, engineering human cells to synthesize their own vitamin C would prevent scurvy from becoming an issue for these future trav-elers. Further, if we reactivated this gene, perhaps oranges and limes would only be used for recreational purposes (e.g., in Martian cocktails) instead of to survive.

Even though this "reactivation" of vitamin C synthesis has been demonstrated in a mouse model, just adding vitamin C synthesis back in could lead to unexpected and potentially negative results in humans due to pleiotropy. There may be a long-standing evolutionary reason that vitamin C was lost as a functional gene from our genome. One reason could be simply that our ancestors acquired enough vita-min C in their diet and, therefore, no longer needed the ability to syn-thesize their own, and this active form drifted into an unfunctional state. In this scenario, reintroducing it may be safe and result in our no longer needing it through supplements. Regardless of the intent, adding genes or entire pathways is complicated and can be unpredict-able. However, these edits and their implications will be detailed out and solved through years or even decades of studying their integra-tion under differing backgrounds, disease states, and environments. As with anything in life, the question will eventually come down to: "What is lost for a specific gain?" With enough time, these losses can be minimized and gains maximized. Eventually, it will be possible to create prototrophic (able to make all required amino acids) human cells.

RECREATIONAL ENHANCEMENT

The rapid pace of research, optimization, and implementation of various genetically engineered improvements to genomes that arose from undirected and unguided evolution will, eventually, result in the usage of these technologies for more . . . *fun* abilities. Once normalized, people will likely use genetic enhancements as a recreational activity. However, the ubiquitous access to powerful technology or therapies may lead to their abuse.

A tragic example of therapy abuse began in the early 2000s in the United States, when opioids became cheaply and readily available. The development of these drugs was for purely medicinal purposes at first, as a means to ameliorate the suffering of patients who had just had significant surgery or who were recovering from an intense, short-term injury. Instead of suffering through unbearable pain, patients could take opioids to relieve some of their pain and eventually recover. Yet the stark reality for these patients was that these drugs were highly addictive. To exacerbate this, the drug usage was widespread, with manufacturers concealing just how addictive they truly were. This led to an era in the United States, when—for the first time in decades— life expectancy actually *decreased* (especially for white males). Unlike previous drops in life expectancy that were driven by war, famine, or disease, this was the first-ever *pharmacologically driven* decline in life expectancy. And for the first time in history, American children, especially among white (and rural) populations, faced a shorter prospective life span than that of their parents.

By the year 2019, opioids were still abused and overprescribed, but a variety of mechanisms were put in place to help alleviate the opioid crisis, such as free clinics, patient support, and social infrastructures. Telethons, home visits, and medical outreach were strongly supported by federal funding and coordination. Eventually, communities that had been devastated by the rapid and ever-present access to cheap sedatives got back on track.

This cautionary tale also applies to the recreational use of genetic-enhancement technologies. There may be a time when, as a matter of curiosity, people begin to enhance their genome or epigenome for better enjoyment of various drugs, or the enhancements themselves could be a kind of drug. Given the epigenetic-editing methods described above, people could find themselves in a state where they decide, "I want to turn on these genes for tonight," or "I want these genes active for the summer." And if they act on the decision, this could increase their enjoyment of a drug by a hundredfold while taking half the quantity. Such a recreational deconstruction and rearrangement of chromatin and genetic regulatory landscapes has no precedent in terms of risk. It might be fine, and have little risk, but it also could be a terrifyingly uncontrolled experiment in cellular disruption and regulatory perturbation.

Nonetheless, there is reason to hope. The usage of recreational genetic enhancements may follow a similar trend to that of HIV patients. Initially, being diagnosed with HIV was a death sentence, especially in the early 1980s. But once antiretroviral therapies, nucleobase analog drugs, and immunomodulatory drugs emerged, life expectancy slowly increased. Then, in 2017, a significant milestone was reached and reported by the *Lancet HIV* journal. For the first time since HIV was discovered, the average life expectancy for HIV-positive patients in the United States was *greater* than the life expectancy of the average US population. The big question, asked by scientists and clinicians everywhere, was why?

This spike in improved life expectancy was due to easy, constant, and reliable access to the health-care ecosystem, which caught any medical problems before they became too complex or dangerous to be addressed by normal therapies. A small infection that could lead to sepsis was caught and treated with a cheap antibiotic. An abscess was found and diagnosed before any other tissue damage occurred. The small insults to the body remained small; life expectancy increased.

At this point in the 500-year plan (the year 2250), there should be a continual presence of a population on a space station around Earth, and their life expectancy may actually increase simply through receiving continuous medical care and monitoring. The high level of scrutiny and focus on the slightest medical changes in any stage of a person's expedition, whether living on a space station or traveling to Mars, will increase the odds that small risks remain small.

Further, some risks may simply be lower in flight. For example, it is clear that a spaceship is a different environment, but is it necessarily a more dangerous one than somewhere on Earth? Perhaps not. While Scott Kelly experienced many molecular changes during his year in space, the majority of them returned to normal. Also, some molecules were actually more stable, and changed less, than what would have been expected if he never left Earth. The epigenome and DNA methylation landscape of Scott were less diverged (compared to preflight) than those of Mark after a year in space. This indicates that just living your life on Earth can induce more epigenetic changes (which could be an indicator of stress or even just simply aging) than the highly regulated, controlled, and methodically planned environment of the ISS. This indicates that a very tight, reliable regulation of sleep, food intake, exercise, and engagement with an engineered environment may be helpful for living a longer, healthier life.

The hope is that, by the end of Phase 6, we would have the first human cells that are autotrophic, or at least prototrophic, and are further endowed with new functions which improve the length, and quality, of life. Astronauts on Mars may begin to approach the life expectancy of humans on Earth. The first baby would be born on Mars. And the plans could be finalized for the first human mission to Titan.

9

PHASE 7: TEST A GENERATION SHIP AND SETTLE HARSH WORLDS (2251–2350)

> The only barrier to human development is ignorance, and this is not insurmountable.
>
> —Robert Goddard

Human existence in the year 2250 will be even more different from the year 2000, relative to the differences between the years 2000 and 1750. People living on Mars will have developed entirely new cultures, dialects, products, and even new religions or variations of current religions. For example, a Martian Muslim will need to pray upward toward the dusty sky, since Earth, and therefore Mecca, will sometimes be overhead. Or, when Mecca is below the Martian's feet, the prayer direction to Allah will stay downward, toward the 38 percent gravity floor; perhaps a second Mecca will be built for the new planet.

By Phase 7, we will likely have the first sprouts of a "green Mars," as envisioned by Kim Stanley Robinson, with life thriving on its own, unaware it's on an entirely different planet from which it originated, because we will have engineered it to have this ability. Given the pressure difference of the atmosphere on Mars relative to Earth (100:1), it may be hard to imagine plants freely surviving outside, but it is not impossible. Further, it is even possible that some bacterial strains could survive in the native Martian soil. Plans to stretch far beyond Mars will be put into action as well, including interstellar missions important for the long-term survival of the metaspecies. We would build ships that do not simply last for one mission, or one decade, but for hundreds of years—the first drafts of "generation ships," where more than one generation of humans lives and dies on the same spacecraft.

But first, we have to answer some key questions. Can we send humans to a planet that is light-years away? Would we need to have multiple generations live and die on the same spacecraft, or have them be asleep? And, if so, what are the ethics behind that? Where should we go? Can we make it?

CHALLENGES FOR THE GENERATION SHIP

Until 1992, there had never been direct evidence of a planet found outside our solar system (exoplanet). By 2020, less than thirty years after this first discovery, thousands of additional exoplanets had been discovered. Further, hundreds of these candidate planets are within the "habitable zone," indicating people may be able to live on them. However, to get there, we need a brave crew to leave our solar system, and an even braver intergenerational crew to be born into a mission that, by definition, they could not choose. They would likely never see our solar system as anything more than a bright dot among countless others.

The idea of having multiple generations of humans live and die on the same spacecraft is actually an old one, first described by rocket engineer Robert Goddard in 1918 (in his essay "The Last Migration"). As he began to create rockets that could travel into space, he naturally thought of a craft that would keep going, onward, farther, and eventually reach a new star. In the twenty-first century, DARPA and NASA launched a project called the 100 Year Starship, with the goal of fostering the research and technology needed for interstellar travel by 2100. Several of that group's members also work with the Initiative for Interstellar Studies (i4is), led by Kelvin Long and Robert Swinney. The i4is works on education and research regarding the challenges of interstellar travel and how to prepare humans for long missions.

This concept of a species being liberated from its home planet was captivating to Goddard, but it has also been the dream of sailors and any star-gazer since the beginning of recorded history. Every child staring into the night sky envisions flying through it. But, usually, they want to also return to Earth. At some point, we would need to plan and construct a human-driven mission, society, and city that are constrained to a single spacecraft that would head toward another solar system—never meant to return.

DISTANCE, ENERGY, AND THE MEDIUM

Such a grand mission will need to overcome many challenges, six of the biggest being: (1) distance, (2) energy needed, (3) interstellar medium, (4) biological/psychological risks, (5) identification and preparation of a world to visit, and (6) ethical dilemmas. Stars are obviously very far apart and hard to reach in a reasonable time, at least with twentieth-century propulsion methods. The distances between stars are usually measured in light-years, a light-year being the distance that light travels in a vacuum in one year (9.46×10^{12} km), or in parsecs (the distance at which the mean radius of Earth's orbit subtends an angle of one second of arc), which is 3.26 light-years (3.086×10^{13} km). The closest known star to Earth, Proxima Centauri, is 4.24 light-years away. Although 4.24 light-years is an extremely small distance compared to the scale of the universe, it will take quite some time to get there, since our current spacecraft are not that fast.

For example, when the Apollo astronauts made their way back from the moon in 1969, the capsule moved at 39,897 kilometers per hour (km/h). The *Voyager 1* probe (the first human-made object to leave our solar system—the heliosphere) moved at 62,140 km/h in 2020, which is 1/18,000 the speed of light, or 1.6 times the speed of the Apollo return mission. At this rate, *Voyager 1* would take 73,687 years to reach Proxima Centauri. Further, in 2020, the *Parker* solar probe was the fastest-moving object ever made by humans, clocking in at 692,000 km/h. At this speed, the required time to reach Proxima Centauri drops to 6,617 years, or put another way, it would take roughly 220 human generations just to make the trip to the closest star. The single-spacecraft progression of 220 human generations is no small ask to the original crew and all of their offspring. Estimates from Jean-Marc Salotti have shown that a minimum of 110 people are needed to survive a trip to the closest planet (Mars), and so more people would likely be needed for a longer trip to another star.

The only way to decrease this number would be to move faster, bringing us to our second challenge: the needed energy for propulsion and sustenance. To decrease the amount of time (and generations) it will take to get to the new star, our speed must increase through either burning more fuel or developing new spacecraft with technology orders of magnitude better than that of the twentieth century. Regardless of the technology, the acceleration would need to be generated either by prepackaged (nonrenewable) fuel, collected from the light of stars (which will be more challenging when between stars), leveraged from elements in the universe itself (such as hydrogen in the interstellar medium), and/or slingshotting off of celestial bodies.

Many ideas to improve thrust technology can help refocus this issue. Nuclear-fusion reactions result in less radiation and more efficient energy conversion than other approaches (>1 percent energy conversion in fusion vs. <0.1 percent in fission) and would further result in crafts capable of much greater speeds. Estimates from the British Interplanetary Society (Project Daedalus), NASA, and the US Naval Academy (Project Longshot) indicate that by using a mass of almost 2,000 tons of fuel for nuclear-fusion ships, the trip could be cut down to just forty-five years (~1 generation) by obtaining speeds of 100M km/h. Further, antimatter drives would be even better in terms of reaction efficiency (such as with the proposed Project Valkyrie), but controlling and even creating antimatter is notoriously difficult because most of the universe is *made* of matter.

Yet even if we have addressed the issues caused by distance and energy by having an incredibly fast, fuel-efficient engine, we still have a third problem: tiny fists of the universe waiting to punch us in the face. For example, a grain of sand moving at 90 percent of the speed of light contains enough kinetic energy to transform into a small nuclear bomb (two kilotons of TNT). Given the variable particle sizes that are floating around in space and the extremely high velocities proposed for this mission, potentially catastrophic events can be lurking in space at any moment. Even small micrometeorites or large dust particles could cause significant damage to our generation ships. This, too, will require further engineering to overcome, as the thick shielding we can currently use will not only degrade over time, but would likely be far too heavy. This may be overcome through (1) the creation of lighter polymers, which can be replaced and fixed as needed in flight; (2) extensive long-distance monitoring to identify large objects before impact; or (3) potentially some type of emitted substance from the front of the ship, capable of absorbing the energy or at least slowing particles before contact.

PHYSIOLOGICAL AND PSYCHOLOGICAL RISKS

As exemplified by the Twins Study and additional NASA one-year missions, the crew of a generation ship would also need to address a fourth key issue: the physiological and psychological stress. One way to get around the technological limitation of either increasing the speed of our ships or protecting the ships from colliding with debris is to, instead, slow biology using hibernation or diapause. However, humans who overeat and lie around all day with little movement in a "simulated hibernation" or "bed-rest studies" can run a higher risk of developing type 2 diabetes, obesity, heart disease, and even death. However, bears can do the same during hibernation without any risk. Why? How?

During hibernation, or torpor, bears are nothing short of extraordinary. They maintain a slightly lowered body temperature, their heart slows to five beats per minute, and for months, they essentially do not eat, urinate, or defecate. They barely move, but still maintain their bone density and muscle mass. They have even been observed giving birth and producing milk while hibernating. The bears become almost perfect recyclers of their own waste, which normally would lead to sepsis, toxicity, or death. Part of their hibernation trick seems to come from turning down their sensitivity to insulin by maintaining stable blood-glucose levels. Bears essentially activate an energy-saving, "smart heart" mode, which uses two of its four chambers and converts their blood into a thick BBQ-like sauce.

In 2019, a seminal study lead by Joanna Kelly at Washington State University revealed striking gene-expression changes in bears during hibernation. They used the same Illumina RNA-sequencing technology as used in NASA's Twins Study to examine the grizzly bears as they entered hyperphagia (when bears eat massive quantities of food to store energy as fat) and then again during hibernation. They found that tissues across the body had coordinated, dynamic gene-expression changes occurring during hibernation. Though the bears were fast asleep, their fatty tissue was anything but quiet. This tissue showed extensive signs of metabolic activity, including changes in >1,000 genes during hibernation. These "hibernation genes" are prime targets for people who would prefer to wait in stasis on the generation ship than stay awake.

Another biological mechanism that we could utilize on the generation ship is diapause, which enables organisms to delay their own development in order to survive unfavorable environmental conditions (e.g., extreme temperature, drought, or food scarcity). Many moth species, including the Indian meal moth, can start diapause at different developmental stages depending on the environmental signals. If there is no food to eat, like a desert with no plants, it makes sense to wait until there is a better time and the rain of nutrients falls.

Diapause is actually not a rare event; embryonic diapause has been observed occurring in more than 100 mammals. Even after fertilization, some mammalian embryos can decide "to wait." The blastocyst (early embryo), rather than immediately implanting into the uterus, can stay in a state of dormancy, where little or no development takes place. This is somewhat like a rock climber pausing during an ascent, such as when a storm arrives, then examining all of the potential routes they may take and waiting until the storm passes. In diapause, even though the embryo is unattached to the uterine wall, the embryo can wait out a bad situation (e.g., a scarcity of food). Thus, the pregnant mother can remain pregnant for a variable gestational period, in order to await improved environmental conditions. The technology to engage human hibernation or diapause does not exist in the twenty-first century, but could be ready by 2251. The first tests of these ideas would be to get to Mars, likely as a round-trip ticket, and then to Titan, likely as a one-way trip.

The impact of weightlessness, radiation, and mission stress on the muscles, joints, bones, immune system, and eyes of astronauts is not to be underestimated. The physiological and psychological risks of such a mission are especially concerning, given that the majority of models were based on trips that were relatively short and largely protected from radiation by the Earth's magnetosphere, with the most extensive study so far from Captain Scott Kelly's 340-day trip.

Artificial gravity, such as rotating platforms generating ~1g centripetal force, would address many of these issues, though not all. Another major challenge, which has been discussed throughout this book, is radiation. There are a number of ways to try and mitigate this risk, be it shielding around the ship (which will likely have similar issues as shielding protecting from space debris), preemptive medications (actively being studied by NASA in the twenty-first century), frequent temporal monitoring of cfDNA for the early detection of actionable mutations, or cellular and genetic engineering of astronauts to better protect or respond to radiation (e.g., Dsup and TP53, as previously discussed). The best defense against radiation, especially in a long-term mission outside of our solar system, will be through the usage of multiple radiation-protective layers: from the ship, to medications, to cells, all the way down to DNA.

Yet even if the radiation problem is solved, the psychological and cognitive stress that comes from the feeling of isolation and constantly being around the same people also needs to be addressed. For example, imagine you have to work and live with your officemates and family, for your entire life, in the *same building*. While we can select the first generation of astronauts and their mental health very well for a long generation ship mission, their children might not be as well adjusted, and previous data have already shown psychological stress.

Analog missions performed on Earth have shown that after five hundred days in isolation with a small crew (Mars-500 project), most of the relationships were strained or even antagonistic. There are many descriptions of "space madness" appearing in both fiction and nonfiction, but their modeling and association to risk is limited. There is simply no knowledge of how the same crew and its descendent generations would perform in ten or a hundred years, and certainly not over thousands of years. There might be factions, rebellions, or even armed conflict, but this has always been true of humans and, hopefully, will be less so with those on a clear mission. Human history is replete with examples of strife, war, factions, and political backstabbing, but also with examples of cooperation, symbiosis, and shared governance in support of large goals (such as in research stations in Antarctica).

CHOOSING OUR NEW HOME

Before we launch the first-ever generation ships, we will need to solve the fifth challenge and gain a large amount of information about the candidate planets to which we are sending the first settlers. A fast way to this is by sending probes to potential solar systems as quickly as possible, gaining as much detail as possible to ensure ships have what they need before they are launched. Work on such ideas has already begun, as with the Breakthrough Starshot mission proposed by Yuri Milner, Stephen Hawking, and Mark Zuckerberg. The idea is simple enough, and the physics were detailed by Kevin Parkin in 2018. If there were a fleet of extremely light spacecraft that contained miniaturized cameras, navigation gear, communication equipment, navigation tools (thrusters), and a power supply, they could be "beamed" ahead with lasers to accelerate their speed. If each minispacecraft had a "lightsail" targetable by lasers, they could all be sped up to reduce the transit time. Such a "StarChip" could make the journey to Proxima Centauri b (only 4.3 light-years away) in only ~25 years and send back data for us to review, following another 25 years of data transit back to Earth. Then, we would have more information on what may be awaiting a crew if that location were chosen. The idea for this plan is credited to Philip Lubin, who imagined in his article, "A Roadmap to Interstellar Flight," a multikilometer array of adjustable lasers that could focus on the StarChip with a combined power of 100 gigawatts to propel the probes to our nearest known star.

The ideal scenario would be seeding the world in preparation for humans, similar to missions being conducted on Mars in the twenty-first century. If these StarChips work, then they could be used to send microbes to other planets as well as sensors. They certainly have many challenges ahead of them as well, requiring them to survive the trip, decelerate, and then (unlike their predecessor StarChips) land on the new planet—no small feat. However, this travel plan is all within the range of tolerable conditions for known extremophiles on Earth that casually survive extreme temperatures, radiation, and pressure. Even our friends from chapter 4, the tardigrades, have already survived the vacuum of space and may be able to make the trip to the other planet, and we could have other "seed" organisms sent along, too. Such an idea of a "genesis probe" that could seed other planets with Earth-based microbes, first proposed by Claudius Gros in 2016, would obviously violate all current planetary-protection guidelines, but it might also be the best means to prepare a planet for our arrival with some microbes that we would want to be there first. Ideally, this would be done only once robotic probes have conducted an extensive analysis of the planet to decrease the chance of causing harm to any life which may already exist there.

THE ETHICS OF A GENERATION SHIP

These biological, tactical, and psychological issues are of course driven by one key constraint on the generation ship: *the passengers are stuck there*. As such, this issue represents the sixth challenge that must be addressed: the ethical component. What are the ethics of placing an entire group of people on a single spacecraft, with the expectation that they further procreate additional generations of people, on that ship? They would have to live with the knowledge that the ship on which they live, or are born into, is the only world they will ever get to know. Certain social, economic, and cultural infrastructure would need to be built into a generation ship along with recreational activities, such as databases with the entirety of Earth's art—something no human would have enough time to completely get through. Entire virtual reality simulators have been built for actors—called the Holodeck in the *Star Trek* fictional universe—and such a sensory experience room is no longer science fiction. Entire body suits, virtual/augmented reality camera sets, and immersive experience sets have been built during the twenty-first century for recreational purposes on Earth, and these would be essential for the generation ship's crew.

They would need to be able to have not only solo experiences but also group efforts, games, and interactions on this ship. Groups of the crew could play each other in a virtual environment, which would require much less infrastructure than traditional sporting events and equipment do. Video games are not just exploratory and recreational events; they are a technological glue of society. As evidence of this, by 2020 the sales of the video-game industry ($100 billion) were larger than those the music ($16 billion) and film ($50 billion) industries combined.

People need games. Across human history, from the Roman era to the latter part of the twentieth century, tens of thousands of people gathered in stadiums to watch sporting events. By the twenty-first century, tens of thousands of people gathered in the same stadiums to watch a skilled video gamer play an entire virtual competition. For example, the top prize for the Fortnite World Cup, in 2019, was $3 million—larger than that for Wimbledon ($2.98 million), the Indianapolis 500 ($2.53 million), and the Masters Tournament ($1.98 million). The concept of a virtual world that can entertain the masses is easy to imagine for the crew in the twenty-fourth century because it already exists in the twenty-first century. If crew members wanted to smell a flower, or feel the ultraviolet (UV) radiation and heat of a midday Midwestern sun, even this could be arranged.

Some critics of sending spacecraft with humans have argued that, even with infinite games, if an interstellar mission cannot be completed within the lifetime of the crew, then it should not be started at all. Rather, because the technology for propulsion, design of ships, and rocketry (as well as our methods for genome and biological engineering) will all continue to improve, it would be better to wait. It is even possible that if we sent a generation ship to Proxima Centauri b in the year 2500, it would be passed by another spacecraft with more advanced propulsion sent in the year 3000.

This "incessant obsolescence postulate," first framed by Robert Forward in 1996, is compelling as a thought experiment. Things do always seem to get better, and technology has continued to improve in almost all human societies. So how can one know when the right time is? Predicting the future is notoriously difficult.

Any concerns about "unfairness to future spacecraft" is overruled by deontogenic ethics and practicality. A good option should not be the enemy of a perfect one. We can send two ships—the first in 2500 and the second in 3000—not just one. If the new ship catches up to the old one, they would likely be able to assist each other and should plan to do so. Further, this obsolescence concern misses the key risk of waiting too long to act. The extinction we are trying to avoid could occur in that 500-year lag, resulting in the obliteration of all life with no back-up.

But even with all of the advanced entertainment, recreational options, and potential hope of a new, enhanced ship appearing any moment, would the crew still stare out the windows into constant star-filled skies thinking of blue oceans? Or would they perhaps be elated about being the "chosen ones" with an extraordinary opportunity to explore and, quite literally, build a new world? The reality is this ship would be their world, and, for most, it would be the only world they would get to experience.

Yet this limitation of experience is actually not that different from the lives of all humans in history. All humans have been stuck on just one world, looking to the stars and thinking "what if?" This vessel, the Earth, while large and diverse, is still just a single ship with a limited landscape, environment, and resources, wherein everyone up to the twenty-first century lived and died without the choice to leave. A few hundred astronauts have left Earth, temporarily, but they all had to return. We are stuck on one ship already. The generation ship is just a smaller version of the one on which we grew up, and, if done properly, it may even be able to lead to a planet that is *better* than what we inherited.

BIOLOGY AT THE SPEED OF LIGHT

The generation ship missions will be limited by any required resources that cannot be synthesized, or picked up, en route. Limited resources will span at least four things: food production, materials development, therapeutics, and waste reclamation. Updated methods for these resources can be deployed incrementally, in order to enable adaptive integration of biotechnology alongside established nonbiological processes and thereby create increasingly sustainable settlements.

Through continual technological development, humans will eventually become more independent from Earth, operating from the limited resources originally from Earth and any minerals found on that planet or moon. Starting on Mars, humans will be limited by Earth's launch capacity and required to live off the land to make their move sustainable—in what is called in-situ resource utilization (ISRU). ISRU minimizes the need of continual payloads from a source planet, while also enabling independence, broader exploration of the world, and mission flexibility.

A central component of ISRU must be the utilization of Earth's biology. Merging genome and cellular editing technologies with extremophiles' innate, inherited strengths could enable life to flourish on worlds like Mars. Earth's organisms (including humans) represent billions of years worth of "method development" to transform raw materials into complex compounds; they also self-replicate, function in diverse environmental conditions, and propagate stored information (as DNA or RNA).

The idea is not to simply enable extremophiles to live on a different planet and let them be. Rather, as DNA synthesis technologies continue to improve, become cheaper, and more transparent, they will begin to be used in space and even on other planets. Through the use of these systems in bioproduction, it will be possible to, in essence, enable organisms' biology to travel at the speed of light.

Imagine there is a settlement on Mars and tissue damage becomes an issue, but at the same time, a new microorganism gene product is made on Earth, which improves tissue repair. The genetic sequences required for this new product can be transferred via data from Earth to Mars within minutes (three or twenty-two minutes, depending on orbital distance). Upon arrival, the organism's DNA could be printed, and the organism could then be living on Mars and making the needed product. Likewise, if a new species is found on Titan or Mars, assuming the organisms are made from DNA that is similar enough to our own, it could be sequenced, analyzed, and sent back in digital form to Earth for synthesis and study in a less resource-limited setting. It would further enable researchers on other worlds to enhance their adaptation abilities through quickly identifying and incorporating positive alterations.

This idea of a "point-to-point biology" would revolutionize how we think about the process of moving life's information around the stars. This pending state of information transfer and human development represents a technological crucible, driving the development of solutions that will not only help humans on Mars, but also provide feedback that can be used to solve some of Earth's most challenging problems. These hopes are part of ongoing discussions by members of the Mars Society, NASA, and other space agencies.

In 2018, a meeting called "Viriditas" was organized by Shannon Nangle and Mikhail Wolfson to merge expertise from industry, academic, and government leaders who want to ensure we get to Mars, and set up a base, as soon as possible. We sketched out a plan to make the "Green Mars" that appeared in Robinson's books a reality. The Viriditas plan included a clear set of goals for Mars, with those for Titan to follow, addressing native biocapacity, bioproduction, and bioreclamation.

Bioproduction, as stated above, refers to the ability to access and build complex molecular products. Its applications have ranged over millennia from bread, antibiotics, and beer to recombinant insulin, CAR-T cells, and novel biomaterials, such as synthetic spider silk. Bioreclamation, which has been used in urban-waste systems since the 1800s, harnesses cellular metabolism to transform harmful or wasted byproducts into a safer, more useful form. Most people do not think about it, but all their stool, urine, and waste are processed quite well. Similarly, the ISS efficiency at reclamation is actually quite good—it can recycle 93 percent of the liquids it collects, including urine, condensed moisture, and sweat from the crew. The recycling process takes eight days and produces water purer than what most drink on Earth. Further, every bit of moisture can be reclaimed and used, no matter its source. When Scott Kelly was asked if he "drank his own urine" while on the ISS, he famously smiled and replied, "Not only mine."

Ideally, engineered systems for waste reclamation would be deployed within the habitats and would not pose a significant contamination risk. The early-stage mission applications of biotechnology will serve as backup, but as their operation is vetted, they will become the primary source of renewables and synthesis.

FOOD FROM AFAR

Food production is one of the most immediate and essential uses of biotechnology on other planets, the production of which all long-term missions will depend on. In fiction, astronauts are often imagined as eating "gruel," which comprises all the essential amino acids and nutrients needed to survive. This gruel is depicted as looking, tasting, and smelling similar going into the body as it does coming back out. While astronauts of the twenty-first century are offered a variety of flavors, the meals are nonetheless designed for efficiency and leave much to be desired. However, on longer, farther missions, the astronauts' diet should offer variety, palatability, nutrition, and enjoyment to improve psychological well-being. Even the ability to smell and nurture flowering plants will be immensely helpful to improve morale of crews on long missions.

There is limited space on a ship and every inch matters. Urban horticulturists face a similar space limitation, which they overcome through breeding compact plants in racks or small storage containers. What they lack in size, they make up for in productivity. Indoor gardens can operate year-round with optimized light, humidity, and temperature enabling them to produce more harvests than a standard farm. As shown with indoor gardens, the shorter the interval between harvests, the greater the amount of food that can be generated. The lessons can be directly applied to how crops are grown on ships and in new worlds.

Optimizations like this have already appeared in the literature, from Zachary Lippman at Cold Spring Harbor Laboratory and Michael Schatz at Johns Hopkins University. In 2019, Lippman and his team created a new tomato with increased yield and decreased time to harvest by tweaking three genes (Self-Pruning [SP], SP5G, and SIER) that control reproductive growth, flowering times, and stem length. CRISPR-editing these genes caused the plants to become impatient, flowering and producing fruit earlier. These mutations also created shorter stems and extremely compact plants, looking less like normal tomatoes and more like a floral bouquet—making them the perfect Valentine's Day present: nutritional flowers.

Beyond gene-edited, optimized plants such as Lippman's, engineered microbial and vegetal organisms can serve as core supplements to the food supply. For example, fermentative production of flavors, textures, and non-animal-based foods has been widely and quickly adopted in the twenty-first century and could drastically enhance food and its overall enjoyment. Successful deployment on other worlds will require the joint development of organisms and fermenters for the planet of interest. As an example, Martian operations would require microbes to use CO_2 and CH_3OH as their sole carbon sources, and the system must be able to tolerate higher radiation and contaminants. This planetary-specific design criteria would expand to most deployed technologies, as seen with the Mars2020 rover's (Resilience) redesigned wheels to better handle the extremely granular sand of Mars.

Both methanol-utilizing heterotrophic and CO_2-utilizing lithoautotrophic fermentation can be used to complement the crew's diet and serve as the initial demonstration of Martian food production. As established on Earth, fermentation can occur in simple stir tanks with engineered organisms to produce complex carbohydrates and proteins. Several methylotrophic organisms, such as *Methylophilus methylotrophus* and *Pichia pastoris*, have been genetically characterized and industrially optimized and could be deployed for large-scale production. Further, engineering methylotroph genes into *E. coli* can be used to improve flavors, textures, and nutrient production, as shown by Jens Schrader at the Karl-Winnacker Institute. Bioreactors with such organisms can have very high efficiency, with a single 50 m^3 reactor able to produce as much protein as twenty-five acres of soybeans, and only require a few days to produce a harvest. Similarly, lithoautotrophs could be engineered to couple hydrogen oxidation with CO_2 fixation to generate oligosaccharides, proteins, and fatty acids for consumption.

Several well-developed terrestrial examples of algal industrialization exist, such as *Arthrospira platensis* for food or commercial algal biofuels. On Earth, however, the high capital costs of building reactors and supplying high concentrations of CO_2 for optimal production are commercially challenging. Due to the CO_2 rich atmosphere on Mars, this challenge actually *inverts* and becomes an advantage. From the combination of efficient light technology, similar to the previously discussed quantum Dot technology, and cellular engineering, photoautotrophs will be produced to synthesize foods rich in carbohydrates, fatty acids, and proteins on other worlds.

The final goal for food production is to be able to grow plants and materials from the planet directly, ideally with little to no processing. Following photoautotroph and hydroponic plant growth in controlled environments, soil-based cultivation of nutritionally rich terrestrial plants such as soybeans, potatoes, and peanuts could begin. For Mars, specifically, the lower solar irradiation, dust storms, different inorganic nutrient profile relative to Earth, and potential soil toxicants essentially necessitates genetic engineering for terrestrial-like growth. Engineered microbes could be used to condition the regolith for crops by minimizing toxicants, enriching for specific nutrients, and decomposing wastes for fertilization. This would be key to enabling the Martians to survive on their own, as well as lead to a true test bed of the plans for Titan and the generation ships. Further, once the ability to engineer prototrophic humans is solved, these "upgraded" astronauts can be deployed to decrease the required nutritional profiles from produced foods.

RAW MATERIALS

Setting infrastructure on new worlds will require a variety of materials, which will first come from Earth but will eventually need to be synthesized from the world itself. Ideally, most materials would be sourced from highly abundant materials on the new world, like CO_2 on Mars and CH_4 or N_2 on Titan. Further, regolith will likely be leveraged as the base component of building, structures, and anything else possible (e.g., underground homes, laboratories, synthesis chambers, tools, and transportation).

Plastics will likely compose a significant portion of mission materials, as they are extremely versatile. Some plastics can be made chemically with established industrial processes, including ethylene to polyethylene, methanol to olefins, and transparent polymethyl methacrylate.

However, the reclamation of these materials will require energy, housing, and specific infrastructure, such as the semiautomated plastics recycler on the ISS (created by the company called Made In Space), which can fed directly into a 3D printer. Biobased materials will be an important class of source materials and recycling for future missions due to their general "reclaimability" through biological mechanisms (e.g., catalysis and metabolic processing). Notably, Martian plastics could readily be produced from the local CO_2 and then reclaimed by life-support infrastructure. CRISPR-modified and adapted organisms could be deployed to accelerate the process and optimize the growth on Mars.

On Earth, adoption of microbially produced materials is largely limited by their low productivity and high cost compared to petrochemical or farmed animal sources. But on a generation ship or on Mars, there is no other choice. The lack of petrochemicals, energy, and animals will require the use of biological material production. Eventually, products created on other worlds such as "Martian silk" will be transported back to Earth, contributing to the new location's economy even if people only want them because they are seen as unique or rare on another planet.

Some pioneering work on creative, biologically sourced materials came from Dr. Neri Oxman, a designer and professor at MIT's Media Lab. She has created works like the "aguahoja pavilion," a winding, cocoon structure that looks like molded honey and is composed of the most abundant biomolecules on Earth, including cellulose, chitosan, pectin, twigs, tree stumps, coral reefs, and bones. She has shown that these biomaterials can be printed by a robot or aggregated together, molded and formed by water, and then arranged into large-scale, stable structures. These engineered, reclaimed biomaterials can be readily used for purposes as fundamental as building construction or as capricious as artistic enjoyment. Some of her artwork already looks like it came from Mars.

As the need for production volume and efficiency increases, substantial manufacturing infrastructure must be developed that relies significantly on resources of the new world, such as large-scale ore mining, smelting plants, foundries, tools, and even technology. This stage of development will enable truly self-reliant planets and drastically reduce the chance of human, and metaspecies, extinction. The technological development which enables this self-reliance on other planets and moons in our solar system will bridge the missing gaps to launch the generation ships.

THERAPEUTICS AND DRUGS

Any long-term mission requires detailed planning of necessary medications to combat the risks of space, reduce risk of accidents, and maintain crew health. During farther missions, even to Mars, an emergency return or a resupply from Earth will be unlikely or even impossible. Safeguarding the physical and mental health of the crew, therefore, demands a plan for long-term therapeutic self-sufficiency, with ISRU, again, playing an important role for self-reliance.

Drugs in space are common. Across the first twenty ISS expeditions, crew members reported 12.6 medication doses per person, per mission, according to an analysis by Rebecca Blue from the Aerospace Medicine and Vestibular Research Laboratory at the Mayo Clinic in Arizona. During expeditions twenty-one to forty, crew members reported an average of 23.1 medications per person, per mission. The medications spanned a range of drugs, including those for mental stimulation, sleep, rashes, headaches, and congestion.

However, survey data from NASA has shown that astronauts report that the medications were only "somewhat effective" or "not effective" up to 40 percent of the time. Medications for rashes, allergies, and congestion do not seem to work as well in flight. This indicates that there is a significant disruption in how the body processes drugs and medications while in flight, likely from reduced total body water, fluid/lymphatic shifts, and altered renal excretion. Such disruptions will almost certainly pose a challenge for crew members on the initial Martian missions, but could (to a degree) be overcome through improved ships with artificial gravity. The more missions astronauts go on, and the more we monitor them, the better we will be able to address these issues through either re-engineering the therapies themselves, genetic engineering of the astronauts, or engineering the ships and environments.

For early missions and planning for generation ships, the entire catalog of therapeutics by astronauts will likely need to be included in onboard cargo, with substantial radiation shielding to prevent their degradation. However, simply protecting the medications taken from Earth will not be enough for longer missions, since 87 percent of the medications on the ISS have only twenty-four-month shelf lives or less, based on FDA guidelines. This includes therapeutics made through mostly automated bioproduction processes including insulin, opioids, and a significant proportion of antibiotic precursors.

Also, there will be a new and large demand for radioprotective therapeutics to develop and sustain human presence in space. Current NASA estimates indicate that the radiation exposure from just a thirty-month Mars mission may meet or exceed the lifetime radiation limits of some astronauts (chapter 3). As of the twenty-first century, therapies such as filgrastim primarily exist to treat the systemic symptoms of acute radiation syndrome. Yet many natural products are also being extensively examined to see if they can protect against radiation damage or induce cellular repair processes. However, to combat challenges of the bioproduction of these therapies in space or on a new world, genetic and epigenetic engineering could be used to address, or prevent, the damage of radiation such as transient activation of DNA-damage-response pathways or the permanent incorporation of genes from other species (like Dsup).

Finally, as the bioproduction and downstream-processing infrastructure available to Mars and generation-ship missions matures, ISRU for therapeutics bioproduction will serve as a primary source of frequently used therapeutics. A mature pharmaceutical-production pipeline would also enable on-demand or emergency production of specific niche medications from a library of microbial stocks as the need arises. New ones could be sent with "point-to-point biology." Finally, DNA diagnostic and sequencing technologies can identify pathogenic species and even guide therapeutic intervention down to specific DNA base modifications, as my lab showed in 2015 on the Vomit Comet and in 2016 with the NASA Biomolecular Sequencer Mission. It is also likely that novel methods for chemical and biological detection (including sequencing-based) would be developed on Mars and that these could in turn lead to improved applications on Earth.

RECLAMATION

On Earth, anaerobic digestion is mostly utilized for wastewater and sewage treatment, especially with concurrent biogas and fertilizer production. On new worlds, developing waste-reclamation methods will be integral to closing the "life-support loop." Waste recycling will be capable of transforming essential elements (such as nitrogen, phosphorus, and sulfur) into bioavailable forms, which would otherwise be challenging or impossible to extract from the initial environment. For example, NASA currently lists bodily/organic waste, respiratory waste, and material inorganic waste as central "capability gaps," which could be addressed through improved biotechnology. In space, anyone's trash (including bodily waste) is a treasure.

At first, bioreclamation will focus on processing urine, sweat, and condensation, as is done on the ISS. Notably, urine is one of the best inputs, given its high concentrations of needed nutrients and chemical simplicity, and is even more homogeneous than solid waste. Unprocessed urine has been used as an agricultural fertilizer since humans settled the Fertile Crescent, and bioreclamation of urine as a microbial feedstock also shows promise. For example, *A. platensis* production of nutrients and proteins is highly similar when grown on urine feedstock or standard media. However, further engineering is needed to address potential inhibition of culture growth from secondary metabolites, tuning microbial metabolisms to the crew's urine composition from mission-specific diets, and other mission or world-limiting factors.

Once bioreclamation technology advances, reclamation systems could begin to process solid organic waste. This process is more challenging, given the greater heterogeneity of human fecal, food, and plant waste streams. Yet the stability of bacteria in new locations like Mars is unknown; it is assumed that the cold storage would result in them not being active. Regardless, crew stool could pose a safety and contamination risk (as it is on Earth) and will need to be dealt with accordingly. On the ISS, most astronaut stool is just packaged and sent out to burn up in the atmosphere. For those on Earth looking up who think they see a "shooting star," sometimes it is just "shooting stool."

Eventually, engineered microbes will be used to improve the efficiency and capacity of existing reclamation systems on Mars. The utilization of more complex material-waste streams, including organic waste, packaging, and chemicals, may benefit significantly from the metabolic engineering of organisms delivered in a custom and on-demand combination of spatially linked microbial consortia (SLMC). SLMC spatially separates organisms into bioreactor modules optimized for individual species or small communities, while permitting the flow of media through each module, allowing for the maintenance of optimal reaction conditions for each individual organism or community and transfer of intermediate metabolites from one bioreactor to the next.

Simultaneously, metabolic-engineering techniques enable optimization of multiple biodegrading or biosynthetic pathways across organisms and reduce metabolic load on a single-engineered organism. Through the union of genetic engineering and SLMC, we may develop custom microbial communities for "on-demand" bioreclamation and materials production from changing waste streams. Also, as discussed in earlier chapters, the biosynthetic gene clusters (BGCs) from a variety of sources are still being discovered on Earth and could lead to entirely new biochemical processes and methods for controlling and reclaiming waste for ISRU. Perhaps even more exciting are the new BGCs and functions that would emerge from Martian and other planetary systems. Evolution has already given many unique adaptations on Earth, and both the accidental evolution and directed evolution in the later decades will likely yield a bounty of new features of life as we begin to catalog life on other planets and moons and even New Earths. Once the neo-Earths are found and profiled, we can begin to pack our bags.

10

PHASE 8: SETTLE NEW EARTHS (2351–2400)

It is better to solve the right problem the wrong way than to solve the wrong problem the right way.

—Richard Hamming

By 2351, biotechnological and other technological advances will enable us to look ahead to a generation ship that might be close to launching. We will have unprecedented abilities to reprocess, remap, and reclaim any of the organisms and matter from the spacecraft that we will need to survive, based on ISRU work that has been pioneered on Earth, Mars, and the first prototypes of a generation ship. We will have ideally not just one planet to target, but many. Even at the time of this book's publication, there are hundreds from which to choose. By 2400, we will want to target the best candidates (figure 10.1), and we will want to have satellites around as many planets as possible (figure 10.2), not just around Earth. Thus, it is worth a quick summary of what we know now, how we came to know it, and what we can hope for by 2400.

TWIN ENGINES OF DISCOVERY: PLANETS AND GENES

As discussed in chapter 3, the discovery of genes has rapidly expanded with the completion of many genomes and their annotations, and it

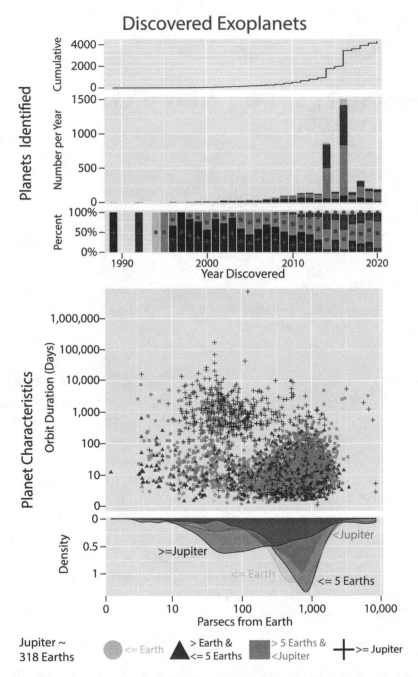

10.1 Discovery rate and types of exoplanets. Each of the >4,000 exoplanets identified since 1989 has added to the catalog of potential worlds we may one day live on. (See color plate 11.)

continues every day. The clear heritability of traits in families and twins has been a persistent observation of humans since recorded time. The quest to map these traits and relate them to specific genes or sets of genes has been ongoing since the work of Mendel and his pea experiments in the 1850s. Just as genetics is the study of complex biological molecules rearranging, reproducing, and creating new variations, astronomy is the study of the interplay of complex molecules and energy (from atomic to galactic) constantly giving birth, dying, and regenerating into what we see as the universe and everything it contains, including the atoms that make up our own DNA.

Since the work of Babylonian observational astronomers in the second millennium BCE, we have known other planets beyond Earth, such as Venus and Mars. Today, anyone with a keen eye can see the Venusian light in the sky at night, and anyone with a decent telescope can find the rings of Saturn as the evening dew sets on blades of grass. While Earth was, for a while, perceived as the center of the universe (the geocentric view), eventually humans realized it was just another planet around a star (the heliocentric view). Over time, we were able to better characterize our system and what it contains, which now hosts eight official planets and many more "dwarf planets," such as Pluto, Eris, and Makemake far out in the Kuiper Belt. But only in the past few decades have we begun to unravel how many other objects orbit the sun (e.g., moons or asteroids). There have been more than 200 moons characterized in our solar system alone.

It was expected, but always uncertain, that other stars in the universe were also dancing with planets of their own. The first discovery of planets outside our solar system (exoplanets) was not made until 1992. The discovery of exoplanets slowly increased after this first discovery, but has since rapidly expanded to include a vast catalog of potential alien homes. Coincidentally, technological advancements in the early 1990s also led to the automation of DNA sequencing, leading to analyses of entire genomes (1995).

The explosion in the discovery of genes within humans and other species represents a beautiful parallel to the rapid discovery of celestial bodies within our own system and the countless other solar systems in the galaxy, twins in a developing womb of knowledge. These paired

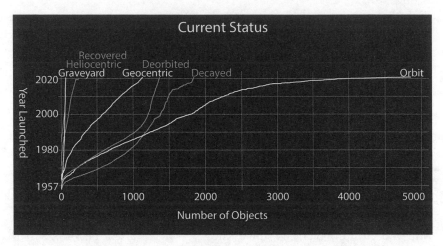

10.1a The number of objects launched into space and their current status. Probes and spacecraft are either still in orbit, decayed from orbit, purposely deorbited, maintained in a geocentric orbit, recovered, orbiting around the sun (heliocentric), or no longer active (graveyard).

engines of scientific discovery are both essential to human survival. They are jointly needed to find, filter, and choose the future homes for humanity as well as to map the biological substrates and biochemical adaptations that we can use to survive on these new homes. We can start with a summary of our planetary options, from worst to best.

HELLSCAPE EXOPLANETS

Imagine if a person aged in reverse. Over time, the person condenses—tighter and tighter—back into a small embryo. This is essentially how planets are born within their planetary disk (essentially their version of an embryonic sack). As you may have visualized, all of the extra material to make a child, spread apart at the beginning, would be much easier to see than a singular cell—this, too, is true for planets (planetary disks). The first evidence of a planet around a distant star (Beta Pictoris) was found in 1984, in the form of a planetary disk (identified in Las Campanas Chilean observatory). However, Earth's observatories are hampered by its fluctuating atmosphere and vibration. To get around

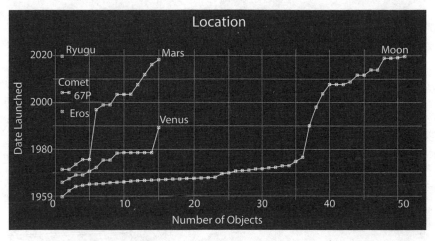

10.1b The location of objects launched into space and beyond Earth. Since 1959, objects have been to the Moon, Venus, Mars, Eros, or Ryugu.

these challenges, NASA launched the Hubble Space Telescope in 1990, giving us the first clear shot to detect distant celestial bodies.

Just a mere two years after Hubble's launch, the first exoplanets were discovered. The discovery contained not only one exoplanet, but two rocky planets revolving around a loud, deadly, constantly irradiating pulsar (a type of neutron star). Though having a pulsar star as its sun would essentially result in the eradication of any life that resembled Earth's, this finding was proof that other worlds *could* exist outside of our own solar system. This meant that any of the stars in the sky could hold one, or many, worlds. Further, any of those worlds could hold any number of moons. And, finally, any of those worlds, or their moons, could potentially hold life.

The next goal was to find a planet orbiting a main sequence star, which is a star in the middle of its lifecycle (like our own sun), rather than dying. In 1995, two astronomers (Didier Queloz and Michel Mayor) found such an exoplanet. However, this exoplanet is more of a failed star than an Earth, or even a Mars. At half the size of Jupiter, with an orbit essentially at the surface of its star, this would not be an ideal place to find life similar to Earth's. Appropriately, these kinds of

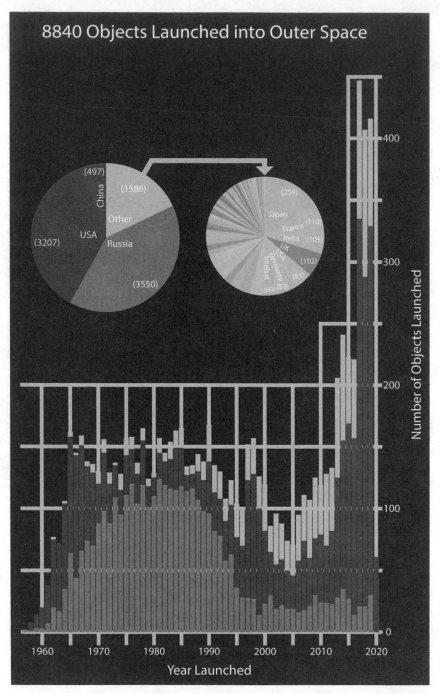

10.2 The number of objects launched into space, summarized by the year of launch and country. This includes Russia, the United States, China, and other countries or agencies, which include Japan, France, India, the United Kingdom (UK), European Space Agency (ESA), Germany, Intelsat, and others. (See color plate 16.)

exoplanets are now called "roasters" due to the intense heat and irradiation they constantly face—like a marshmallow deep in the heat of a campfire for billions of years.

Then, in 1999, a new method for finding exoplanets called "transiting" was tested and confirmed. Transiting occurs when a planet passes in front of a star and casts a shadow. In 1999, teams led by David Charbonneau and Greg Henry found an exoplanet that passed in front of the star HD 209458 in the Pegasus constellation. The transit allowed astronomers not only to detect that a planet was there but also to get a sense of its atmosphere. As light passed through the atmosphere, energy was absorbed and emitted by the planet's thin surface. These light spectra represent energy being absorbed and emitted according to the elements of the periodic table, including elements of water, oxygen, nitrogen, and carbon. But how could they do that?

SEEING ATOMS

To understand the overall spectrum better, it is essential to understand how these spectra came to exist in the first place. Isaac Newton is widely regarded as the founder of spectroscopy (the study of spectra) because, in his first work on using prisms (the treatise *Opticks* in 1704), he described with great detail how a white light is composed of separate elements of color. These separate elements compose the same rainbow of colors that we all see in the sunlight when it is passing through the rain. Newton observed that the prism was not creating new colors or modifying the light to change it; rather, the prism seemed to be separating constituent parts of what was previously thought to be "pure" white light. An idea was born. What if spectrum might be a way to classify what is present inside the light? But a simple glass prism did not provide much detail beyond the fact that there were component parts that needed separating.

Then, in 1821, a young Bavarian named Joseph von Fraunhofer had the key idea of using a "diffraction grating," instead of a glass prism, as the source of light dispersion. This grating was made of wires that were closely aligned in a parallel grate, similar to the tight lines of cells in the feathers of birds, and that separated out higher-resolution colors from various light sources, such as the sun in the sky or a candle. Indeed, the idea for this method allegedly came from Fraunhofer holding a bird's feather up to the sunlight, a beautiful and simple experiment that anyone can replicate. Upon using this diffraction grating, Fraunhofer created a quantified "wavelength scale" and then published extensive observations on the sun's light, carefully noting which bands of color were consistently present. But he also noted that some wavelengths of light were consistently missing—dark bands. These bands of "missing" wavelengths were a mystery at the time, but were suspected to be additional evidence of the fragmented and discrete "chunkiness" of light. To this day, these dark bands are known as "Fraunhofer lines" in honor of his work.

By the mid-1800s, it was becoming evident that each element of the periodic table, and even each molecule, might have very specific emission spectra as its core response to absorbing light. Then, in the 1860s, William and Margaret Huggins applied spectroscopic techniques and methods to the stars for the first time. They hypothesized that if the universe was composed of the same atoms as those on Earth, then the same spectra could be used to discern the elements present on any observed planet or star. Indeed, this hypothesis proved to be true, and they were the first to take a spectrum of a planetary nebula and to distinguish nebulae from galaxies based on their spectra—specifically for the aptly named Cat's Eye Nebula, which looks like a feline wink from a faraway galaxy.

These and related observations inspired the work of Niels Bohr and others in the early 1900s, which confirmed that not only were the spectra predictable at specific wavelengths, but they could solve another problem in quantum mechanics: How did the atoms' quantum states exist? Bohr focused much of his work on the hydrogen atom, the most abundant in the universe and also the simplest atomic structure (one proton, one electron). He proposed that light coming from pure hydrogen might be a consequence of the atom's absorbing energy at a very specific frequency and then emitting it at a specific wavelength—which would explain the Fraunhofer lines of dark gaps.

He was right. As atoms absorb energy, their electrons shift to a higher orbital plane and then emit light as they drop to a lower atomic orbital. Upon this drop, energy is released (often in the form of light, as per the equation $E = h\upsilon$). Because these emission spectra are consistent across all labs tested on Earth, the assumption is that the spectra observed from a planet or star in another galaxy can reveal the chemical elements present. Indeed, these spectra enable a discrete map of the elements present in any sample in a lab or any surface of a planet transiting in front of a distant star. The transiting method captures spectra emissions of a given exoplanet as it passes directly between its star and the observer, just like Fraunhofer holding a feather up to the sun. This feat was the manifestation of a 350-year-old dream, begun by Newton, whereby an instrument orbiting Earth used the visible emission spectra from another planet to discern the elements present.

PURGATORY PLANETS

Once it was clear that spectral mapping and transiting methods worked, discoveries came fast. In 2001, with the use of Hubble's spectrometer, the original exoplanet HD 209458 b's actual atmosphere was observed, the first evidence of atmosphere outside our own solar system. In 2005, the Spitzer Space Telescope used infrared wavelengths to observe two exoplanets at once. In 2007, the Spitzer telescope was now capable of identifying information on larger molecules, as opposed to just individual atoms. Further, astronomers David Charbonneau and Heather Knutson used Spitzer to generate the first weather report of an exoplanet, including cloud cover temperature. It was not clear if this alien world would be a nice place for a vacation, but at least we knew what clothes to pack. But nothing is quite as good as seeing a planet with your own eyes. In 2008, the first direct image of an exoplanet was taken, the astronomical equivalent of a school picture. Fomalhaut b was found using Hubble, nestled a relatively close twenty-five light-years away. However, this planetary child was morbidly obese—about three times the size of Jupiter.

Given all the success shown with the transit method, it was time to bring out the heavy hardware to find more exoplanets. In 2009, the *Kepler* spacecraft was launched to hunt for more transiting planets as they moved in front of their home stars. The hunt did not take long.

In 2011, *Kepler* found its first rocky exoplanet, the smallest planet yet found, which is basically a large chunk of iron about 4.5 times the mass of Earth. However, like many worlds identified before it, this planet was very close to its sun—even closer than our own Mercury. It is basically another scorcher planet that would not be kind to any life-form we know.

HEAVENLY GOLDILOCKS EXOPLANETS

With the continued advancement of methods and Kepler's unsleeping, unblinking eye peering into the universe, we finally got lucky. In 2014, we found an exoplanet (Kepler-186f) that resembled Earth, dancing in the "Goldilocks" zone. If you've never heard of classic fairy tale "Goldilocks and the Three Bears" (for example, if you are an alien reading this book), it tells a story of a child who passed over porridge that was "too hot," then another that was "too cold," until finding one that was "just right" to eat. In terms of humanity's quest for good exoplanets, a "Goldilocks" planet is one that is in a similar life-enabling range to its star as Earth's (potential liquid water), based on both the distance from the star and information on the star itself, such as the temperature.

Kepler-186f is slightly larger than Earth, likely rocky, and within the Goldilocks zone, thus enabling liquid water to exist. Even though it is 500 light-years away, Kepler-186f is a planet that future humans could one day journey to—the first putative "neo-Earth." Then, in 2015, a second, bigger neo-Earth was found, called Kepler-452b. This planet is 1.6 times the size of Earth, features a 385-day orbit around its star, and it, too, is present in the Goldilocks zone.

By 2016, the *Kepler* mission had created a catalog of about 1,200 exoplanets, about 40 percent of them with a composition similar to our own Earth. Within this catalog exist many more Goldilocks planets. There's a system that's only forty light-years away and straight out of our wildest, utopian sci-fi dreams, containing seven Earth-sized planets, many of which may have liquid water on them (TRAPPIST-1 system). Another Goldilocks exoplanet also exists around our closest neighboring star, Proxima Centauri. At only 4.2 light-years away, we have a potentially habitable planet we could realistically send probes to. This planet is one on which a human could not only hold liquid water in their hands, but do it while standing safely on the surface. The universe suddenly feels smaller.

Given all these emerging discoveries, in 2018, NASA launched the Transiting Exoplanet Survey Satellite (TESS) to further accelerate exoplanet efforts. Its goal is to find planets ranging from small rocky worlds to bulging gas giants. If all goes according to the plan, TESS is expected to find some 20,000 exoplanets during its mission, including as many as 17,000 planets larger than Neptune, as well as an estimated 500 Earth-size planets that may, someday, be visited by humans (even for vacation). There is an ongoing digital counter at JPL that shows how many total planets have been found and how many are in the habitable zone. By the twenty-first century, the Exoplanet Archive (housed at Caltech) showed >4,200 total exoplanets, with >360 in the habitable zone. But this begs the further questions: How may total planets are there? Is there life somewhere on them?

DRAKE, FERMI, AND THE GREAT FILTER

An early formulation of the likelihood of life in the universe was first proposed by Frank Drake in 1961, now known as the eponymous Drake equation. This equation estimates the amount of life capable of communication in the universe, with the idea that—if life exists—it might have the same technology as we do, and we can hear it:

$$N = R_* * f_p * n_e * f_l * f_i * f_c * L$$

From this equation, the number of civilizations in the Milky Way galaxy (N) with electromagnetic emissions that would be detectable by human technology in the twenty-first century is a function of the rate of formation of stars that might create planets where life could emerge (R_*), the fraction of those stars with planetary systems (f_p), the number of planets, per solar system, with an environment suitable for life (n_e), the fraction of suitable planets on which life actually appears (f_l), the fraction of planets where intelligent life emerges (f_i), the fraction of civilizations that develop a technology that releases detectable signs of their existence into space (f_c), and the length of time such civilizations release detectable signals into space (L).

Many of the early estimates from this equation were nakedly speculative, considering the equation was created before the first exoplanet was even discovered. However, updates to the Drake equation in the twenty-first century resulted in some scientists stating that it is almost a certainty that life exists somewhere in the universe. Current estimates posit that there are about 1.5–3 relevant (life-supporting) stars formed per year, with each star likely having at least one planet, and many of these planets possibly supporting life—either in the Goldilocks zone or with the chance for life to emerge at the edge of life's features. From these numbers, and assuming both that the number of planets where life can emerge is 1/10,000 and that the number of planets with intelligent life is 1/1,000,000, with 5 percent of those intelligent life-forms being able to communicate, and life having lived at least a billion years, then there are 3.9 million life-forms we should have heard from by now.

So where are they all? This same question led Enrico Fermi, the Italian-American physicist, to declare this a "paradox" in 1950. If true, the very high estimates for the probability of such life is at direct odds with the staggering lack of any evidence for this extraterrestrial life. It could be that intelligent life-forms like humans are extremely rare, or that intelligent life and civilizations are short-lived due to war or limited resources (called the *Great Filter*), or that they exist in some other dimension of communication we cannot currently measure. Also, the numbers could just be too optimistic. By adjusting only a few of these numbers to be more conservative (by a factor of 10 or 1,000), it is suddenly easy to estimate that there is only a 0.000001 percent chance of aliens existing from which we could hear any signal.

SEAGER, BIOSIGNATURES, AND SPECTRA

The other solution to this question is to get a better equation more focused on what we know about life. In 2013, Dr. Sara Seager, a leading planetary physicist at MIT, proposed an update that maps out a parallel equation that can work next to the original Drake equation. The "Seager equation" is built to estimate the odds of detecting signs of life on exoplanets by their telltale biosignature gases. It posits that the number of planets with detectable biosignature gases (N) is equal to:

$$N = N^* * F_Q * F_{HZ} * F_O * F_L * F_S$$

N^* is the number of stars within the sample, F_Q is the fraction of quiet stars, F_{HZ} is the fraction with rocky planets in the habitable zone, F_O is the fraction of observable systems, F_L is the fraction with life, and F_S is the fraction with detectable spectroscopic signatures of biosignature gases. With this new equation, an appropriate sense of the products of life has been infused into the search for such life.

Although some of these terms are derived from related work in planetary science and the Drake equation, some are new, such as the biosignature gases and degree of "quietness" of a star, which discerns how much radiation would be stemming from the system. Active stars often have high levels of UV radiation, which drives a rapid degradation of most known biosignature gases, and thus makes life harder to find in such a system. From the first check of these numbers in 2013, there were perhaps a total of two planets expected so far (including Earth)—thus, not a big number. Yet, as more and more planets are found and these numbers' accuracy increases, the reliability of these measurements will also increase. Further, since so much of what we know about exoplanets comes from the very, very small sliver of the galaxy that we've examined, these numbers are hard to reliably calculate in the early twenty-first century.

The TESS and the James Webb Space Telescope (JWST) will undoubtedly identify more exoplanets, many of which will be within the habitable zone, and some of which will be a putative neo-Earth. Also, the number of biosignature gases could increase. Currently, if an alien species was looking at Earth, oxygen would be one of the tell-tale signs of a clear kind of biochemical or biological process for its continual generation. This is because, in its native state, oxygen is quite rare in the universe. Since it holds a high electronegative potential (the chemical tendency to "steal" electrons from other atoms), it does not last long in most environments. To see a lot of oxygen on a planet, such as the very high 21 percent that is present on Earth, it must be generated on a continual basis by some chemical or biological process.

Moreover, it is not just the presence of specific gases that is key, but their amounts and patterns. On Earth, carbon dioxide levels rise and fall with each revolution around the sun, as the Northern Hemisphere's greater surface area and base of photosynthesis (especially across Asia and North America) absorbs more CO_2, and then it lowers again for the latter half of the year—every year. The planet takes a big breath, inhaling CO_2 in an annual, predictable, and tell-tale sign of the variegated density of life on the planet and continual photosynthetic response to the stellar output.

Seeing similar dynamics on other exoplanets would also be cause for excitement and lead to their being deemed top-priority sites for generation ships. Where signatures are detected, new ships could be pointed in their direction to enable an extraordinary era of guided evolution that continually learns from interstellar discoveries and, potentially, entirely new intelligent species with an entirely different history. The only way to find out what is truly possible is to leave our first home and explore.

THE CASE FOR TITAN

It is no longer a question of *whether* we can find exoplanets, or *if* Earth-life planets might exist outside of the one we know, but instead *which ones* we should target for future homes. Similarly, in genetics, it is no longer a question of *whether* we can modify genes, or even *if* we should modify genes to cure diseases (as it is already being successfully done), but instead *which ones* we should engineer, modify, or synthesize. Many efforts from 2000 to 2300 will focus on setting up sustainable, self-reliant cities on Mars. But this will merely be humanity's version of going to college. Leaving the house you grew up in, traveling just out of the reach of your parent's ability to instantly help you, and testing your limits, boundaries, and potential—all while having fun, learning a lot, and likely getting into trouble. But all of these lessons will be learned, whether we like it or not, and will help us continue to move forward into other worlds in our solar system, with the first human mission to

Titan (one of Saturn's many moons), ideally occurring in 2300, and with preliminary settlers by 2400. This idea may not be as impossible at it sounds, with many planetary scientists noting the planet's benefits, as Charles Wohlforth and Amanda Hendrix did in their book *Beyond Earth: Our Path to a New Home in the Planets*.

A moon is essentially just a planet with altered loyalty and cooler views. As such, there is no fundamental reason that life cannot exist on a moon. In fact, moons can be even larger than planets (e.g., Titan is larger than Mercury). Titan even has a thick atmosphere—about 50 percent denser than Earth's. This enhanced atmosphere will even protect surface dwellers against radiation from GCRs and HZE particles. Further, it has stable, liquid lakes on its surface comprised of methane and ethane—as if an oiler's fantasy theme park has come to life.

One differentiating quality of Titan relative to Mars or Earth's moon is the readily sourced energy. The abundant hydrocarbons can be a ubiquitous source of building materials, fuel, and even plastics. Through supplemental nuclear power and oxygen from engineered plants and algae, it would even be possible to perform electrolysis of water. Titan's intense atmosphere and winds can come in handy here, yet again, to produce energy through wind turbines.

Other abundant resources on Titan could offer additional sources of power from the well-established reaction of hydrogenation of acetylene ($3H_2 + C_2H_2$). Acetylene is used extensively on Earth for industrial processes, including welding for building large structures (because it can create a flame at 3,600K); illuminating portable lights and lighthouses with carbide lamps (which burn acetylene); and acting as a substrate for semihydrogenation (to ethylene), which can then be used to create polyethylene plastics. Finally, acetylene can be chemically converted to acrylic acids, which are the basis for vinyl, acrylics, some glass, paints, resins, and many plastic polymers. We could build an Earth-like home on Titan by pulling atoms from the atmosphere and have vinyl records made to play Titanian music.

Another attractive quality of Titan is how mobile we would become. We would gain advantages both as individuals, bouncing around the world, and also as explorers, with the moon becoming the solar system's rest area to fuel up, grab snacks, and stretch your legs before journeying to other stars. Titan's gravity is a breezy 14 percent of Earth's, making transit even easier than on Earth's moon. Further, it moves slower, relative to Earth, making vacations (or more practically launches) easier to plan. Titan takes sixteen Earth days to complete an orbit around Saturn, twenty-nine Earth years to revolve around the Sun, and has a similar tilt to Saturn and orbits in its equatorial plane resulting in similar, seven-year seasons. Imagine a warm, fun summer that spans seven years, where you can stare at Saturn's rings the whole time.

In the scheme of planetary dating, Titan isn't exactly perfect, but who is? Its largest "red flag" would be that it is emotionally unavailable. Basically, it is cold. However, assuming the cold could either be made more tolerable or at least kept at bay (similar to Antarctica) with physical protection, people could survive on its surface. The only spacesuits needed for going outside would be ones that provide oxygen and retain heat—there is no need for pressure to be maintained or extra protection from radiation in the suit because the atmosphere has those covered.

Of course, Titan is not ready for life yet. It has such a dim light (a hundredfold less sunlight intensity reaching it than Earth) that Titanian crops would be hard to create using only natural sunlight. Given its longer days, the circadian rhythm of plants and algae would need to adjust. But, as noted before, qDot lighting or other sources of artificial light could readily be made to power the growth of algae and Titanian food (not to be confused with titanium food, which has little flavor). Titan also does not yet have breathable oxygen, which (so far) humans seem to enjoy. And, though stated before but not to be underestimated, it is cold—very cold, at 94K (vs. Earth's 290K average). Nonetheless, given that Titan sits at the outer edge of the solar system, it would be a monumental step toward the expansion, preservation, and exploration of Earth's life.

The twin engines of discovery—planets and genetics—enable us to identify, design, and engineer our way into a stable presence of Earth's life on harsh worlds. But how do we decide what organisms to send, what needs to be engineered into them, and how this engineering should be done?

GENETIC GUARDIANS

Assuming exoplanet and genetic discoveries continue at the same pace until 2351, we could have as many as 5 million exoplanets from which to choose for settling. Also, assuming that genome discovery, engineering, and mapping continues at an exponential pace, we could have millions of genomes and genetic constructs from which to plan a genetic payload. This is where a decision gets a bit tricky—who do we send? Humans are currently unique for their forward-projecting conscious states, which give us our sense of duty to the stewardship of all life. So we are the obvious choice. But what other organisms might have that sense of consciousness or could develop it soon?

The goal is to maximize the long-term preservation, survival, and even enjoyment of life for species we send to new worlds—potentially including species which have at some point become extinct. But how do we choose from the diverse, amazing life on Earth? To make this decision, we can add a fourth layer (the Guardians) to three other components of an ecosystem: (1) the producers, (2) the consumers, and (3) the decomposers. Simply put, Guardians are the protectors of planets' ecosystems.

For a new planet, producers will come first. Producers use water, air, and sunlight to make their own food and food energy; every planet will need them. They include microorganisms and plants that aid in energy production, nutrient processing, and recycling wastes. These organisms will also enable rapid experimentation on new worlds—including genetic optimization. Also, the producers can be highly mission-specific, with (for example) Earth's extremophiles selected based on the new world's innate extremes. The producers could also come from our "microbial astronauts" launched on light, fast probe-ships to seed worlds of interest. When possible, once a world is stable and capable, other consumer and decomposer species can be sent. Finally, the genetic Guardians (like humans) would be sent, who would serve as architects and engineers of the new diversity and variation of life that would inevitably emerge. This could either happen by generation ships or, ideally, through point-to-point biology that continually writes species' needed DNA with the functions appropriate for a new planet.

As of the twenty-first century, humans are the only species capable of conjuring the forward-projecting conscious states required to design such a mission and react to any unpredicted hurdle which may come up. This is the innate ability that gives us our sense of duty for the stewardship of all life. But the fact that this is unique to humans is really an accident of history. As there are many species capable of handling extreme temperatures, there's no reason to think only humans are capable of a conscious state that enables the awareness of the possible extinction of all life.

Once a world has been established to support Earth's life, it's entirely possible that we, then, leave the new world in the hands (or paws) of a new Guardian species. Without the constraint and pressure from humans, additional species may be able to develop their usage of tools, language, and consciousness. This not only enhances the overall preservation of life through diversifying the amount of species capable of acting as Guardians, but further gives additional species more ability and control of their lives, which is more deontologically ethical. This therefore makes the classification of a Guardian, or any other ecological state of a species (producer, consumer, or decomposer), dependent upon the current abilities or limitations of that species at a given time and on a given planet.

The initial generation ships will require humans as the Guardians. However, these first ships would ideally also contain other species capable of becoming Guardians. Through directed evolution and the aid of humans, as opposed to our abandonment, additional Guardian species may be able to emerge even more quickly on these first worlds.

Which species would fall best into this "pre-Guardian" category? Given that humans are primates, we may as well include some of them for the trip. Pygmy primates have the benefit of a small surface area and could survive in small areas of a ship and an alien world, although they are not huge fans of extreme temperatures. Cats and dogs might be good candidates as well, given their current integration into human society and companionship. Dogs, foxes, and horses are some of the best examples of directed evolution, through selectively choosing desired traits. Some of these traits can even be successfully selected for in a relatively short period of time (tens to hundreds of years).

SUCCESSFUL DIRECTED EVOLUTION

The most notable example of directed mammalian evolution was the "silver fox" experiments in Russian Siberia from 1959 to the early 2010s, led by Dmitri Belyaev and Lyudmila Trut. The goal of this experiment was to create a doglike phenotype from wild silver foxes, relatives of "red foxes" (*Vulpes vulpes*), through creating a "tameness score" and breeding the top 10 percent from each generation. Any foxes that showed traits of biting, growling, and anger were separated from the calm, emotive, and compassionate foxes, which were bred, thereby resulting in a human-guided evolutionary process.

Within only six generations, what looked like a new species had already emerged, with floppy ears, curly tails, and a 50 percent reduction in glucocorticoid levels (a stress hormone). The newly docile foxes began to exhibit behaviors just like those of dogs, including licking the researchers' hands, snuggling when they were picked up, and wagging their tails when the researchers approached. The chemistry and physiology of the fifteenth-generation foxes were even more distinct, including smaller adrenal glands; increased serotonin levels, mottled "muttlike" fur patterns; shorter, rounder, more doglike snouts; and doglike body shapes (chunkier limbs). Perhaps the most eerie development from this directed-evolutionary experiment is that the foxes began to follow human gaze and interact as any other pet would, watching and waiting for a tasty snack as if they have been by our side for hundreds or thousands of years.

Amazingly, selecting for the tameness score *alone* was enough to create this domesticated, highly doglike phenotype. When the researchers (including Lenore Pipes from our laboratory and Andrew Clark at Cornell) looked at the selected foxes' DNA and RNA, they found a gene that likely created the tame behavior: SorCS1. This gene synthesizes the main trafficking protein for glutamate receptors in the brain and implicated enrichment for synaptic plasticity in the foxes. Even more strikingly, when the foxes died, the brains were examined for genetic and functional differences. The RNA and gene expression profiles (when compared to those of the earlier foxes) exhibited changes in their brains' transcriptome profiles—including some aspects that resembled those of modern dogs. Specifically, 146 genes in the prefrontal cortex and 33 genes in the basal forebrain were differentially expressed, with enrichments in neurodevelopmental pathways, as well as in the serotonin and glutamate receptor pathways. Within sixty years, the aggressive traits of wild silver foxes had *almost entirely* been removed from their genome, chemistry, brain, and overall behavior. Dogs were made all over again—all within the life span of an average human.

These same ideas could, in principle, be applied to any other species with high-level cognitive abilities. There are many great candidates on Earth, including dolphins, which already have an intelligence estimated to rival that of humans, blue whales, or other cetaceans. Ideally, we repeat the silver-fox breeding experiments, this time selecting for intelligence. This would not only improve these organisms' chance to survive on a new world, but also elucidate intelligence substrates that could be genetically utilized across different species.

By the late 2300s, these directed-evolution methods will likely have been used in a number of different species, resulting in smarter and more resilient organisms better equipped to live on new worlds. This process will create a positive feedback loop, learning from the adaptations of life on all worlds, including from the first "native Martian," who is born from parents who were also born on the red planet, and becoming more adaptive to multiplanetary living over time. Genomic and physiological changes seen just in the first generation of native Martians could yield clues for helping others survive on all other planets.

Future humans will look different, assuming we survive long enough to continue to change. The nuchal ligament, which attaches to the back of the skull and is needed to keep the head straight while running, would perhaps become less pronounced in residents of lower-gravity planets, while those who have adapted to higher gravity may have this as well as other muscles and ligaments more pronounced in their necks. Native Martians could readily be imagined as playing with directed-evolution pets, such as those with increased radio-resistance and modified ectoderms, thriving in the reduced atmosphere and dim Martian light with blue sunsets. Entire colonies of the engineered organisms could even be frolicking with native Martians in the dust.

But this will not end with Mars. It may even be possible to someday have a multiplanetary, engineered genome which is consistently being improved upon and added to as we travel to and find new worlds. Triggers in the environment of Titan's atmosphere may release the cascade of the Titan-specific gene package, whereas being on Mars would only enable the Martian gene package. This would enable even more genetic freedom and living abilities such that a given family could be spread across multiple planets and easily (with appropriate technology) visit one another.

MATTER-AGNOSTIC TOWARD COGNITION

Each of these goals for highly engineered and intelligent species creates a possible risk that they would overtake humans as the "dominant" species on that planet or in that system or someday even the galaxy. That might be fine, if they continue to plan ahead and feel the same deontogenic drive toward the universe, but there is also a chance that they may object to the deontogenic ethics described here or even find a wiser path forward.

The pros and cons of diversifying the Guardian duties apply not only to biological life, but to artificial-intelligence (AI) platforms and robots as well. Despite the warnings from Hollywood movies where machines have taken over Earth and humans have been subjected or exterminated, the benefit of the robotic and AI platforms is that they can assume many of the human tasks and much of the leadership that can enable us to survive on alien worlds. With this in mind, AI—and not biological life—may in fact become the second Guardian sentience. Indeed, some of the harshest places in the universe will likely only be *able* to be visited by machines, and there are situations where these mechanical life-forms may be better suited than biological. There is already a rich history of machine exploration within our own solar system, such as the *Cassini* probe or the *Curiosity* rover.

To put this in context, consider again Peter Singer's philosophical work. He argued that humans who chose to eat meat "because they were just animals" were actually enacting a form a racism at the species level, or "species-ism," in 1975. The will to ignore the capacity for suffering in other animals is not enough of a reason to just allow humans to eat anything they want to eat. These animals at least deserve our moral consideration, just as a human being—even with severe brain damage—would deserve consideration to avoid suffering. Regardless of species, all life with a long-term memory, a capacity for suffering, and awareness of pain deserves to have that considered and ideally reduced. Just as being racist has been found to be an abhorrent, fallacious viewpoint, someday perhaps the same will occur for speciesism.

But, what if "matter-ism" is the next logical and moral fallacy that we cannot see today? If robots or artificial species that we have engineered or that come to exist take over our role as Guardians, we could not declare that "these pieces of matter are not the same as us, and thus can be disregarded." Using the same syllogism as above, it is likely that we cannot morally discriminate against matter (carbon-based or otherwise) either.

Even the mechanical life of the future, if truly sentient and aware of extinction, would have a prerogative and self-sustained imperative to preserve its own, and others', existence. For it, too, "existence precedes essence." Thus, at this point in the ten-phase plan, we would have no preference for organic versus inorganic matter as the stewards and protectors of the sentience of the universe. The Guardian species by then may all be mechanical in nature, or biological, or even a combination of both within a singular being. At this point, all existing Guardians will hopefully look out to the horizon and feel that sense of duty to all self-replicating and sentient entities across the universe. For any Guardians to survive, they must work together and expand throughout the solar system (and universe) and support each other.

Once this vision is achieved, the final stages of preparation for becoming interstellar entities can begin. Self-reliant, mobile cities can now be engineered, packaged, and sent to the best candidate planets to ensure our—or our inheritors'—survival. For the first time in the history of life, life itself will choose its own sun to orbit.

11

PHASE 9: LAUNCH TOWARD THE SECOND SUN (2401–2500)

If we stay here, we'll all change. The air. Don't you smell it? Something in the air. A Martian virus, maybe; some seed, or a pollen.

—Ray Bradbury, "Dark They Were, and Golden-Eyed"

The full, multisystem backup plan of Earth's life will initiate with the launch of the first generation ship around 2401. Beyond prolonging the existence of Earth's life, this initiative may even lead to the identification of new life unlike anything we have ever seen (similar to unnatural base pairs, UBPs, discussed in Phase 5). Through the identification of new findings on different worlds, we can look, again, into the universe with new lenses—such as new biosignatures and emission spectra in distant stars and planets. This will be the beginning of a sustainable presence of life in the universe, one that isn't at the mercy of the collapse of a single system (be it rogue asteroids, solar output, or ecosystem collapse) and, further, not defined by the limited window of life on only one planet.

However, every time humans have expanded into a new place, a drive for independence has erupted and often led to bloodshed, suffering, and death (e.g., the American Revolutionary War). Given this history, it is likely that a new settlement, colony, or outpost of humanity would eventually want its own rights, power, and governance— and could we blame them? The society that leaves Earth to settle new worlds must eventually be as independent socioeconomically as they are with the production of their own required resources. They must not be exploited as a means to gain remote resources.

However, these societies shouldn't be cut off from the rest of life, either—that would only hinder the advancement of technological, biological, and ethical discoveries. Trade and autonomy between systems must be planned for, expected, and engineered into its initial design. Goods may come in digital form, such as point-to-point biology trade between ships, planets, and stars, as well as the creation of rare physical goods derived and created in new locations. Ideally, there will be a rich exchange of ideas, products, and knowledge between systems fulfilling the deontogenic duty as humans (or subsequent form of sentience) across a whole society.

A DEONTOGENIC SOCIETY

The potential deontogenic society of 2401 will feature the end of planned-obsolescence products (e.g., car parts purposefully engineered to fail after a year), enhance the widespread use of durable and efficient multipurpose machines, and optimized processes spanning across biology, genomics, and medicine. Rather than a simple, short-term capitalist drive to maximize dollars per unit on cheap devices, the new economic drive will be focused instead on maximization over the long term, including long-lasting products and investments that last generations. This will not only be a more ethical way to operate, but will actually be required to operate in a multiplanet state, where technology must last for long trips and with limited or no repair capacity. In space, there is no instant gratification or next-day shipping to replace something.

To adapt to this new era, companies will have to be built to last for the long term, with the best companies having multigenerational track records of reliable research and durable products capable of handling extremes across multiple planets. While companies that lasted for hundreds of years were rare in the twenty-first century on Earth, there were some examples. For example, Kongō Gumi, a manufacturing company in Japan, has been running since 578 CE. Similarly, experiments will be planned that span hundreds, thousands, and eventually even tens of thousands of years, enabling the study of complex heritability of loci selected through directed evolution, as well as planetary-scale terraforming.

By 2401, the majority (or potentially all) of genetic disease will no longer cause human suffering. DNA sequencers (readers) and synthesizers (writers) will be both accurate and common. Technologies will enable the seamless mixing and synthesis of genetic networks in easy, accurate, and cheap ways, akin to DJs mixing and making music with a beat sampler. As noted by music critic Nate Harrison, both the sampler and the turntable were key tools largely responsible for the birth and development of hip hop. With the sampler, any sound that could be recorded could be used as part of a new composition.

The DNA sequencer is the "sampler" of genetics; the DNA synthesizer is the turntable. This means that any DNA fragment can be sequenced, and then used as part of a context of genome design. With the turntable, musicians can mix and match components to create new musical compositions. With the synthesizer, any DNA fragment can be combined to create new genetic compositions.

In this era, humans have the ability to control their underlying genetic code, controlling for how their molecules fundamentally change in response to stimuli and enabling new abilities. This will enable an unprecedented ability to build, edit, and transplant cross-kingdom combinations of genomes, which we will need to survive on new worlds. The majority of other diseases or causes of human death on Earth will also largely decrease, including those associated with birth, cancer, and even accident. These advances will go beyond the Guardian species and be implemented to improve the quality of life for other animals as well. There will no longer be an excuse for animal suffering, caused by humans or even parasites. The continual engineering of animals, plants, and single cell organisms will enable them to thrive in previously unimaginable conditions.

UBIQUITOUS AND CONTINUAL BIOLOGICAL ENGINEERING

The merger of two key events will fundamentally change human society forever, both of which are under active investigation with promising results in the twenty-first century. First, the successful engineering, clinical validation, and broad utilization of exowombs. This process will start slowly, initially used only to finish the development of children born prematurely. Over time, these processes will improve, the fundamental laws of human development will be uncovered, and these technologies will be used earlier in human development. Eventually, they will be capable of bringing an embryo to a fully developed and perfectly healthy baby. Second, cellular engineering (including genome and epigenome editing) will continue to improve over time, enabling the editing, incorporation, or removal of any level of cell biology in accurate, inexpensive, and easy ways. Once these have been vetted within in vitro therapies, they will then be tested as adult in vivo applications and, eventually, within embryos.

However, the safest and most ethical way to incorporate these technologies within an embryo or fetus will be before it is developing within an exowomb, as opposed to its mother (or *one* of its mothers as previously discussed). In this way, the developing baby can be constantly monitored, and if/when issues arise, they can be quickly addressed and fixed. Eventually, it may be *more* ethical to have your child develop within an exowomb rather than its native mom, for both the health of the woman and the child. Reproduction would be further decoupled from sex, as new technologies to "engineer out" diseases will be developed, and highly specific, largely automated, monitoring technologies will be developed to further ensure the birth of a healthy child.

Moving from the embryo to the adult, here, too, the number of genetically engineered people will continue to rise over time (figure 11.1). This process already started in the twentieth century through the usage of ex vivo derived autologous (using the patient's own cells) and allogenic (using donor cells) cellular therapies, as described in chapters 3 and 4 (e.g., CAR-Ts). These technologies have been further under active development in the twenty-first century to derive "universal" cellular therapies, which can be picked out of a freezer and used to treat anyone regardless of their genetic background (HLA type) without the fear of graft versus host disease (GVHD). These universal cells will, for the first time, enable the widespread usage of cellular therapies by overcoming three of their main hurdles simultaneously: (1) the economic

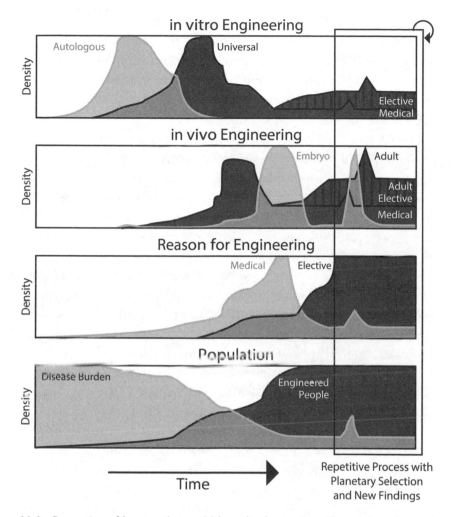

11.1 Proportion of humans that could be edited over time. The number of genetically engineered people (top) would increase first for therapeutic purposes and then
for enhancement reasons, until it would become unethical to not do any edits, and then
another round of edits would appear that would reduce the disease incidence a second
time (bottom) within a repetitive process.

burden of deriving a product of each person, every time they need it;
(2) the amount of time it takes to derive a patient's own therapy; and
(3) the unpredictability of how a patient's cells will respond to the
engineering or act once engineered. Finally, these genetic engineering

therapies could be integrated into nanoparticles with engineered tropism to enable highly specific and precise engineering within a given person. These universal cells and in vivo engineering capabilities will work in synergy, creating customized cells to address more diseases for more people. These will eventually become the standard of care, leading to more cures and fewer treatments.

Indeed, assuming exowombs are shown to be safer than carrying a child full-term, it would be considered "poor parenting" to not include exowombs within a human's pregnancy plan, assuming you had the means. Or, future mothers might be viewed as "romantic" for doing gestation "the old-fashioned way." Further, once it has been shown that genetic editing can be done safely and improve long-term health, it would actually be recommended by geneticists and pediatricians where appropriate. As such, to *not use* it would be akin to refusing to use vaccines in the twentieth century and, appropriately, result in scorn and derision of the parents. This idea is just an extension of the same "herd-immunity effect" that is granted with widespread vaccination efforts, except now applied to heritable diseases.

The convenience of an exowomb and the empowerment for women notwithstanding, these will give us the greatest chance to develop healthy babies and even adults with lower medical risks. Through the efforts of The Online Mendelian Inheritance in Man (OMIM) database (even though it is a misnomer; diseases affect women too), which annotates genetic disorders, and ClinVar, the clinical variation database, a road map could be followed to correct mutations associated with single-gene disorders (e.g., cystic fibrosis), multifactorial disorders (e.g., diabetes and asthma), and even risk of infections (e.g., malaria).

However, this type of engineering would not only be applicable for familiar genetic diseases such as Down's syndrome, but would also improve immune, circulatory, and neurological systems. As an example, an optimized immune plan could be integrated to improve antigen-presentation and cellular differentiation to better identify foreign or unwanted pathogens while reducing the incidence of the destruction of one's own cells, as commonly happens in autoimmune and rheumatoid diseases. Further, OMIM contains over 25,000 entries as of 2021, some of which are genetic variations that lead to "abnormal laboratory test values" (such as dysalbuminemic euthyroidal hyperthyroxinemia). These, too, could be improved.

THE GENE TRUST

The widespread application of these genetic therapies, and their adoption into society, will eventually blur the distinction between enhancements and therapies—from the first cell onward. The use of some genetic engineering strategies will likely require development in exowombs, while at the same time people will choose to do more engineering because they are already planning to use an exowomb—driving an increased usage and demand for both technologies.

There would likely be campaigns that promote the most, best, and healthiest options for embryos, using slogans like "From the first cell you get, to the first cell you choose." Eventually, *not performing* cellular optimization and *not using* exowombs would only lead to increased risks and endangerment for the child, their to-be mothers, members of the society, and even future generations. Performing cellular optimizations would be perceived as a duty that spans parental, social, intergenerational, and deontogenic dimensions, and fulfilling this duty will transform communities, countries, planets, and eventually all of humanity.

However, as is the case with any major update to code, in this case our shared genetic code, creating a backup is strongly encouraged. A "genetic backup" of unedited humans will likely exist, living in a separate society of their own, based on groups of some people's disdain for change across human history. However, given enough technological progress, this society is not a requirement for our backup. If necessary, we could simply undo our editing, reverting back to a previous version in our genetic library, using the same engineering technology—an even easier feat than the de-extinction of entire species. There is even an example of this in the twenty-first century, the Svalbard Global Seed Vault "Crop Trust" in Norway. Storing data and cells across human generations would result in a similar "Gene Trust," further maximizing our chance of long-term survival and protecting against extinction caused either by our own hubris or by accident.

UPRIGHT GENETICS

The idea of selecting and modifying entire human populations, while controversial and the basis of horrific Nazi experiments, has actually already been deployed successfully for ensuring health. The simplest example of this has already occurred in the twenty-first century, in the Ashkenazic and Sephardic Jewish communities. These religious groups (and especially Orthodox Jews) usually marry and have children only within their religion. Because of this selectivity, there has been a historically high rate of several genetic disorders, including Tay-Sachs disease, which is both painful and fatal. Tay-Sachs symptoms often appear in the first year of life as a loss of motor skills, seizures, vision and hearing loss, and muscle weakness. In almost all cases, when children carry both copies of the defective gene, they die by the age of four. There is no treatment or cure.

In 1983, Josef Ekstein, a rabbi in Brooklyn, New York, had an idea to stop the spread of these diseases: identify a potential couple's risk *before* they even meet or have children. In modern genetics, this is called "carrier screening," where potential parents have their genetic risk factors examined to reduce the risk of passing on a disorder to their would-be children. If potential mates were identified as being "carriers" for a set of diseases, they could avoid children with other carriers. This idea would decrease the incidence of Tay-Sachs and other diseases *without* the need for abortion, IVF, or PGD—an important requirement for the Orthodox Jews, given the opposition of such elective treatments, their complications, and expense in the 1980s.

Rabbi Ekstein started a project called *Dor Yeshorim*, meaning "upright generation," taken from an Old Testament verse (Psalms 112:2). This service set up genetic testing to identify if an individual carried mutations that could cause Tay-Sachs disease and saved this information into a database. People were then given a random code which they could use to call a number and find a mate, so that their offspring would be unlikely to have the disease. Over time, these systems became more sophisticated, and by 2021, one could simply order a relatively inexpensive ($225) test to check for a number of different diseases, including cystic fibrosis and Canavan disease. Further panels may even be added depending on background risks.

Did love hit you hard and fast? Good news, there's even an "emergency test" for just this case ($450). As an extraordinary demonstration of the power that genetic planning can bring, there was a 90 percent reduction in the incidence of Tay-Sachs in ultraorthodox Brooklyn Jewish community after implementing this approach. Due to the widespread uptake of these approaches, a previously haunting and awful death sentence for children has been mostly removed from the world's Jewish population.

Indeed, the Dor Yeshorim project has been so successful that it has moved on to addressing other diseases. The "standard panel" of genetic testing for Dor Yeshorim now includes cystic fibrosis, Canavan, Niemann-Pick types A and B, familial dysautonomia, Fanconi's anemia, glycogen storage type 1A, Bloom syndrome, mucolipidosis type 4, and spinal muscular atrophy, among other diseases. Some of the testing is also specific to the Jewish background of each person, and there are specific gene panels for each possible genetic mixture. For example, Sephardic Jews are tested for twenty-three monogenic diseases, whereas Ashkenazi Jews are tested for ten.

The incredible decrease of devastating diseases has not gone unnoticed. Inspired by these stories, additional services have emerged to enable broader access to carrier screening technologies. One such example is a dating app called Digid8, spun out of George Church's lab, which enables people to share their genetic data and match with others, to decrease the chance that their offspring will inherit any number of severe diseases. If such a method were implemented for all severe diseases, there could be a dramatic reduction in disease burden and suffering around the world.

As genetic engineering treatments and enhancements become more common and broadly adopted, there will likely also be a strong social drive for genetic selection, similar to the carrier-screening methods of the twenty-first century. However, this selection process would have to be balanced by continuously examining and weighing the risk and reward associated with any genetic, epigenetic, or cellular alteration.

There will further need to be a clear characterization of what actually constitutes a *disease* or, more appropriately, once enhancement and therapies begin to be blurred, a *limitation*.

INDIVIDUAL, INTERGENERATIONAL, AND INTERPLANETARY RISK SCORING

Genetic risk factors of the twentieth century were driven by ancestry and historical patterns of human migration. For example, cystic fibrosis is the most common autosomal recessive disease among people of Northern European heritage, but is less common in Asian populations. Similarly, sickle cell anemia is more common in people with sub-Saharan African ancestry, but less common elsewhere in world. As with carrier-screening technologies, different methods could be implemented to decrease the overall global burden and suffering caused by these diseases through informing individuals of their to-be children's risk susceptibility. Over time, as these burdens decrease, their ancestral associations would also change, resulting in continually altered risk profiles.

As such, the population- and genetic-mapping tools employed in 2400 for disease or phenotype likelihood will need to continually reassess the associated risks and rewards of interventions to diseases, limitations, and enhancements. As such, a combined score could be derived for a specific disease with a specific treatment, in a specific location (including the current or pending planet), which weighs the impact these factors have for an individual, society, and economy. This score would then enable the comparison of one treatment to another for a specific disease or the severity of one disease with its best-possible treatment against another disease, thereby focusing research efforts to improve overall well-being.

Such a score did not exist in the twentieth century, since it depended on many factors that were hard to quantify. However, much like the Drake equation, whose formulation was useful as a guide to laying out the driving factors related to the question of intelligent life in the universe (vs. the exact answer), a similar genetic quotient of risk across one's life can be a useful guide and thought experiment. For such a metric, the elements' impact and relevance would change over time, as society, technology, and treatments would continue to develop. Such factors would need to include those which weigh on the inflicted individual, those around them, the current population, and the future impact that a given treatment (or lack of one) could have on the economy and society, which would all be planet specific.

Indeed, at least thirteen factors would comprise such a "Lifetime Risk Score" for given phenotype (figure 11.2). These factors include:

1. The "background" population life expectancy;
2. Age of symptom onset;
3. The expected age of death;
4. Treatment success likelihood;
5. The overall quality of life (1 means no decrease in the quality of life, 0 absolute suffering or death);
6. Proxy suffering (suffering of loved ones or those around them, causing a negative feedback loop on patient's suffering);
7. The function of onset (relative to background population and as such location specific),
8. Penetrance and the risk to develop the disease (1 if the patient already shows symptoms);
9. Pleiotropy (1 no pleiotropy, >1 complex negative phenotypic associations, <1 if associated treatment helps other phenotypes, such as comorbidities);
10. Balancing selection (1 if no evidence of balancing selection, >1 if evidence of selection, which may confer worse outcomes elsewhere, <1 if positive outcomes);
11. Heritability of condition;
12. Anticipation (the altered severity of disease over generations (1 if evidence of no worsening);
13. The economic burden (taking into account the cost of treatment, supportive care, and prevalence for current and future generations).

The values associated with this risk score (S) will depend on three things: the disease (d) or condition being examined; the treatment (t), or lack of one; and the place (p) of context, including new planets (p). To put this into a formula, this risk (S_{dtp}) could be represented as shown in figure 11.2, as the aggregate score of combined disease-treatment-place risk (S_{dtp}).

The equation is designed such that a larger relative value would result in a worse outcome and therefore something that would need to be addressed. As an example, the earlier the onset relative to the background-population life expectancy, and the less time the person is expected to live, the larger this score will scale (age: A). The treatment success would be the chance that the treatment will work for the individual, such that a lower chance of success would result in a higher overall risk score (success: S). As treatment success tends toward zero, the overall score will approach infinity—making the usage of the treatment highly unlikely, unless it is the only option and results in a better outcome than no treatment at all.

Similar to treatment success, quality of life will further move this score, such that if quality of life is extremely low, approaching zero, the overall score will approach infinity. Proxy suffering further increases the score when a patient's quality of life is not maximized, expected age of death is not that of the background population, and there are others around who may suffer. This may then further increase the suffering of the individual, seeing their loved ones suffer on their behalf, which would need to be accounted for (quality: Q). However, if the quality of life is maximized and expected age of death is equal to background-population life expectancy, then proxy suffering would be equal to one, resulting in Q equal to one.

If a treatment is capable of extending the expected age of death beyond that of the background-population, then the age term (A)

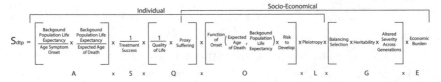

11.2 This score allows for the comparison of how a disease, condition, or enhancement risk changes depending on treatment and location. This further allows for the relative ranking of the severity of a disease given limitations at the time, which can then direct future research.

will decrease. However, the overall well-being of the individual would depend on the amount of additional suffering they may experience in their additional time. As such, the overall score must be a function of the onset of the disease; for example, if their disease will get extremely worse for the majority of the additional time they have acquired from the treatment or location, then they may actually be worse off. The output from this function of onset is then further weighted by the risk to actually develop the disease or condition (onset: O), which is the "penetrance" of the allele—the proportion of people who get the disease if carrying a risk allele. If the risk to develop is close to zero for the individual, then the overall score would be low (approaching an asymptote of zero), whereas this term would be equal to one if symptoms already exist.

The overall pleiotropy will further play into both the individual and generational risks as well as the complexity of treating the condition (L, pLeiotropy). This value would be one if no pleiotropy is identified, and could actually be less than one in cases where there are overall positive associations or the treatment reduces comorbidities. Similarly, many effects which span generations would need to be tracked and analyzed (G, generational value), such as balancing selection, which may further result in additional pleiotropic effects. Also, if/how a condition may worsen over generations (anticipation), as with Huntington's disease, would be a key attribute that will likely strongly favor gene editing therapies which remove diseases from the population. Finally, economic burdens (E) would be a major factor including both the monetary and social strain of not only the therapy, but the continuation of a therapy across generations, if not corrected.

Such a score has the ability to simply compare the impact of one therapy versus another on a given disease or phenotypes, for a given person, in a given society. But beyond that, these scores could inform someone on how their life might change depending on where they live—whether on Earth or on different worlds. As an example, say a disease only starts to present itself at 80 years of age, but has a function of onset that rapidly causes a poor quality of life and death within a year (making the expected age of death 81). This disease would have a much lower risk if the background-population life expectancy was only 50 (say, on board a ship), relative to a location with an average life expectancy of 100 (say, on future Earth), given that the person may likely never reach the age where symptoms develop. Therefore, the risk score of this disease, even if given the same treatment options at both locations, will be dramatically lower for the individual if they lived on the ship relative to Earth. The same idea can be applied to positive evolutionary selection pressures on one planet which can serve as risk factors on another. As such, these values will need to be analyzed and readily available for any combinations of interventions across different planets.

As seen with carrier-screening, it is likely that the first wave of genome engineering decisions based on S_{dtp} would start in a singular community, to address a given disease, which causes a large amount of burden and suffering. Once overall benefit is realized, these techniques would then be broadened to additional populations, and eventually span the globe for a multitude of ailments. The number of genetically engineered people would reach its peak, and the incidence of disease would fall (see figure 11.1).

Then, in vivo, embryonic engineering would replace the usage of somatic engineering, since it will decrease the economic burden, heritability, and any potential increased severity of the disease over generations, while offering the same improvements in quality of life. It is also likely that a "burst" of more subtle diseases or phenotypes would emerge after widespread adoption due to our own ill-defined understanding of the diseases and alleles, or even from new diseases that we face, as we gain the ability to explore new locations and worlds.

Once we pass the "postediting burst" of cryptic diseases, most editing would then be for adult cells, because the previous embryonic editing would be inherited and embryos would only need to be further edited for newly emerging or previously missed diseases. This could not only resolve the genetic risks for many diseases and disorders, but could lead to a world where any trait, from any organism, can be "toggled" on and off, as needed or simply wanted (such as recreational, low-oxygen living on the top of Mount Everest).

THE DANGER OF ABLEISM

However, this ostensibly beneficial plan to get rid of "bad genes" has several issues. First, as the extremely long yet oversimplified equation reveals, it will be no simple feat. While laudable in application to eliminate diseases, such as Tay-Sachs, it is this very same rationale that was used to lay the foundation of eugenics in the early 1900s. Eugenics has been the folly of many geneticists, including "councils of experts" in the 1920s who championed their enthusiasm to avoid disease and to imagine a world without suffering. They quickly started to think about mapping and quantifying "underperforming" traits beyond just severe diseases. The idea would be to essentially to put "chlorine in the gene pool" and remove any trait deemed "undesirable." Then, as now, there are councils of geneticists who annotate, quantify, and stratify diseases based on their own measures of severity, creating well-meaning metrics like S_{dtp}.

Yet, this ideology has historically had a devastating effect on society and the field of genetics. The eugenics movement of the twentieth century and its geneticists ended up eviscerating the liberty of individuals rather than safeguarding it, especially reproductive liberty. Laws were drafted and passed across the United States that required forced sterilization of any member of society who possessed a low intelligence quotient (IQ) or was deemed "retarded" by metrics from the *Diagnostics and Statistical Manual of Mental Disorders* (*DSM*). Such IQ metrics led to an estimated >70,000 sterilizations in the United States and the theft of the ability of those individuals' ability to have their own children. Importantly, the *DSM* itself changes over time, which means that "disease" and our definition is not just a matter of pure empiricism, but can be (and almost always is) influenced by society at the time, as discussed above. For example, homosexuality was listed as a disease in 1952 in the first edition of the *DSM*, but the listing was removed in the second edition in 1973, and has not returned since.

Another issue with the goal of removing "bad genes" is that it can promote "ableism," which is the prejudice or discrimination against individuals with disabilities. Work from Jacqueline Mae Wallis, among others in disability studies, has shown that some reject the idea that deafness is a disability because it can enable an entire world of experiences, culture, and language that is unique and distinct from that of "the hearing." Moreover, deafness can also lead to new sensitivities of touch and tactile "feeling," especially for subtle vibrations, that may not readily be available for hearing people.

The unique physiological, cultural, social, or cognitive abilities that people experience due to their specific genetics, whether it is a disability or not, have led some couples to express an interest in editing or selecting IVF embryos so that their children will have similar experiences—including some deaf couples wanting to ensure that their child is deaf. In a 2008 survey taken in the United States, Susannah Baruch and colleagues found that 3 percent of IVF–PGD clinics provided PGD to couples who wanted to use it to select an embryo for the *presence* of a "disease or disability," such as deafness or dwarfism.

The selection of any trait, whether for or against it, is complicated. Removing one *disability* may also accidentally remove an *ability*—as is often the case with pleiotropy and complex traits. Just as removing CCR5 to resist HIV also increased the risk of West Nile virus infection, tampering with one gene and one phenotype can increase the risk of exacerbating others. Imperfections in human knowledge and historical errors in the application of genetics to "fix" diseases are often at odds with the disability-rights movement, which aims to secure equal opportunities and rights for all people with disabilities, rather than removing these aspects of their identities.

Many people with disabilities view themselves as, and are, functional, active, and vibrant members of their communities, because of—*not in spite of*—their disabilities. While self-assessment of one's own abilities and status can certainly be a complicated and imperfect process, there are obvious challenges with assuming that another group of nondisabled (or differently abled) people can judge the internal states of others. A prominent researcher in the field, Elizabeth Barnes, noted quite correctly in *The Minority Body* that "the intuitions of the privileged majority don't have a particularly good track record as reliable guides to how we should think about the minority, especially when the minority is a victim of stigma and prejudice."

Indeed, the imperfect use of this genetic knowledge leads some to conclude that widespread carrier screening or genome engineering to avoid disease should not be done at all. Instead, some argue we should stick to the status quo and more "natural" methods of mate selection and procreation. However, this argument unnecessarily limits human-kind's ability to safeguard life, reduce suffering, and to fulfill our deon-togenic duty as Guardians. The goal is not, and should not, be the removal of liberty or freedom, but instead the addition to it.

Whenever given two options, there are always two more: both and neither. On a new planet, and also on Earth, it may be good to have the best of both options. We can actually have both the benefits that may come with a previously designed disease (e.g., enhanced senses), as well as the benefits of not having that disease (additional abilities or not having to live in constant pain). The overall goal, rather than forcing mutually exclusive choices, should be to drive technologies and safe methods that can create the greatest amount of cellular, reproductive, and planetary liberty, including the ability to not only choose a world to visit, but to be able to thrive once there (figure 11.3).

PLANETARY AND CELLULAR LIBERTY

So how can there possibly be a "both" option when discussing removal or selection of traits? This ability will likely depend on the specific traits, themselves, but it can be possible. Instead of only acquiring new abili-ties, such as enhanced vibration detection for the hearing impaired, at the expense of others, such as hearing, it could be possible to engineer both abilities within the same person. This question can be addressed by the S_{dtp} score, to determine how it would impact someone's quality of life, potential proxy suffering, and overall economic burden. Again, these values will be both time and location dependent. Within space, it may make more sense to have a better ability to sense vibration, com-municate without the need to hear, and even think in more abstract ways. The ability to selectively turn on genes, redirect cells, and recreate tissues of interest within the body can have a profound impact on the places to which a body can travel and the degree of self-control over one's own cells. Some level of this cellular liberty is, to a degree, already

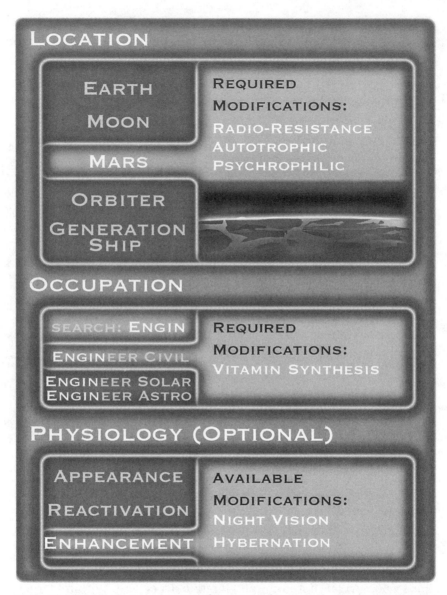

11.3 The Genetic Programming Interface. In the future, both jobs and recreational work could employ a paradigm for editing certain genes, pathways, and networks, as needed.

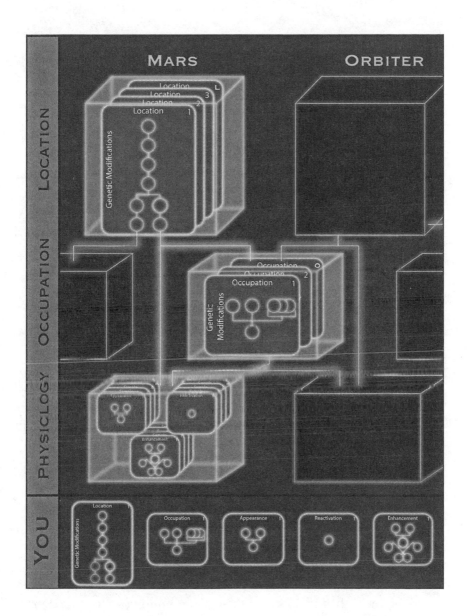

present in the twenty-first century in the form of sex-reassignment therapies and surgeries.

But will every "ability" that accompanies a disease or condition be engineerable? At first, the answer is likely no. For example, hardship from a disease or the experiences gained from it are not purely negative—going through such hardship can, in turn, strengthen the individual. It can give them drive to then want to be a part of how the world operates. Such resolve can cause them to work toward solving a missing link in science that is required to make the generation ship a reality, but one that was only born out of the struggle with an ailment.

Many people living with such a disease, even those without hypersensitive abilities, may say that, even if they were given the chance to go back in time and never have the disease, they would not. However, this is typically for one of two reasons: (1) one cannot go back in time, so why crave an impossible state, or (2) they truly like who they have become and know a large part of what made them who they are is what they experienced (including the disease). But the reality is—there will be hardships in other forms, through unrequited love, the loss of a family member, trying your hardest but still not succeeding. There is no evidence that having any given disease is the causative reason they are resilient—as opposed to people who already were, or would become, resilient anyway. As Guardians, we should minimize the suffering of all subsequent species and individuals while maximizing their liberty and knowledge. While it may not be possible, at least at first, to engineer the resilience and optimism that can come from dealing with fighting a disease, it undoubtedly can be acquired through other means.

One such means to create the inspiration of resilience could, again, be an individual's choice to "turn a disease on," such as wanting to feel a brief, targeted destruction of a tissue or the pounding and simultaneous stabbing of their brain in response to light during a migraine. Even though this ability would take time to develop, it could also potentially be acquired through advanced virtual reality—experiencing an entire life in the shoes of hardship, all while, in reality, being unknowingly safe in bed.

Perhaps most importantly, there is one thing that this deontogenic society and advanced cellular and planetary liberty will in fact remove:

helplessness. The dread of a pending death sentence from birth, at least for most diseases, could vanish. Moreover, the new technologies can enable altering of previously immutable, genetic traits, giving a malleable path to a new sense of belonging or role. This belonging will no longer be forced—it will be chosen. Some may choose to live on Mars, others may choose to turn off their hearing, and others may choose completely different traits; all get to choose their traits and features.

Ideally, the best combination of genes and traits would be selected for each specific exoplanet, and then integrated into the genomes for long-term use. Here, too, there is a question of how much cellular or personal liberty is gained or lost by predefining the traits of a person going to a specific world or on a specific mission. Denotogenic ethics would argue, however, that the package of edits that enables the most people to survive on the greatest number of worlds would be the ideal option for "preloading" the human genome and surviving as long as possible. Even economically, this would make the most sense, as it would minimize the number of elective operations an individual may want. The best option for any embryo, anywhere, is that which enables the greatest cellular and planetary liberty.

EMBRYOS IN SPACE

The main challenge of "spacefaring embryos" is their safe development. It is simply not clear whether embryogenesis will function for human embryos in space as well as it has on Earth. As of 2020, no embryo has gone through the entire process of fertilization, development, and birth while in space. It is perhaps worrisome that several mice studies have shown that having just a part of fetal development take place in space can disrupt the vestibular system's development, causing issues when they were born on Earth. All of Earth's life has depended on, and developed under, the pressure of 1g gravity. The lack of gravity in space, on the ISS, and on potential ships to Mars, is the primary reason for this developmental concern.

However, this is by no means impossible to address. The answer is further engineering, although not necessarily genetic. If there is a rotating section of the spacecraft or generation ship that creates 1g, or even just partial gravity, it is likely that the embryos would develop just fine. However, any mechanical issues relating to this section of the ship pausing for any period of time could also be detrimental to the developing embryo. Medical interventions leveraging the previously discussed mechanisms of diapause could be used here to give the mother (or exowomb) and her developing child more time.

But there is at least some evidence that the early stages of embryogenesis can successfully occur in space. In 2016, a Chinese research team used a microgravity satellite to send 6,000 mouse embryos into space to observe their gastrulation, taking photos every four hours of development. The team, led by Duan Enkui, Professor of the Institute of Zoology at the Chinese Academy of Sciences, found that most of the embryos developed into blastocysts, indicating that this early stage of development is possible in space (at least for mice).

If space embryogenesis turns out to be too complicated, there are many other ideas which could be put into action, many of which require large improvements to twenty-first-century technology (genetic and mechanical), as previously discussed. One such idea (first proposed by Adam Crowl in "Embryo Space Colonization to Overcome the Interstellar Time Distance Bottleneck") would be to send frozen embryos directly to an exoplanet of interest. Once a spacecraft arrives at (or near) the new planet, automated robotics, artificial intelligence, and artificial uteruses (exowombs) would be deployed to create, raise, and teach the new human beings. Most of the technology for autonomous raising of children does not yet exist—much to the chagrin of tired parents—so it remains to be seen if this would be a viable childrearing method. Thus, it is likely that some people, potentially multiple generations, would need to be alive on the ship to aid in its navigation as it heads toward a new star.

PLANETARY DATING APPS

Even after the extensive planning and analyses conducted on Earth, and even with the aid of probes, the exoplanet we set off toward may not fit our needs as perfectly as we had hoped. This may force the ship and its crew to switch to their plan B, and further possibly even plan

C, and so on. To aid this contingency plan, generation ships would ideally be sent to multiplanet systems, containing multiple potentially habitable planets or, at least, multiple potential moons to live on. The uncertainty of when a mission will actually end, and where the crew might one day call home, further underscores the need for generation ships to be as self-reliant as possible.

Before we can start to think about what system we want to target, we first need to be able to score and rank potential exoplanets based on their similarity to Earth and their overall habitability. Both metrics received a specific formulation in 2011 from Dirk Schulze-Makuch and colleagues, published in *Astrobiology*. They proposed a two-tiered classification system for exoplanets, the Earth Similarity Index (ESI) and the Planetary Habitability Index (PHI). The ESI is composed of both an "interior ESI" for a planet—including its density and radius—and a "surface" ESI, based on the likely surface temperature and escape velocity. The PHI is built upon the presence of a "stable substrate," or place to land, available energy, life-related chemistry, and the potential for holding a liquid solvent. Though the PHI may give a more accurate representation of what planet would be best for the habitability of known life, the ESI offers an estimate using more readily available metrics:

$$ESI = \prod_{i=1}^{n}\left(1 - \frac{x_i - x_{i0}}{x_i + x_{i0}}\right)^{\frac{w_i}{n}}$$

Where x_i and x_{i0} are specific properties of the extraterrestrial body and of Earth, respectively, w_i is the weighted exponent of each property and n is the total number of properties being measured. The PHI is similar to the Bray-Curtis Similarity Index, which is also used in microbiome studies of diversity. This index constrains the planet's similarity score between 0 and 1, with identical to Earth (based on all analyzed metrics) being 1, and the most divergent from Earth possible (in terms of ESI) being 0. This metric can further be expanded to include additional features of a planet (such as using mass, radius, escape velocity, flux, density, and temperature all together) and even simplified by allowing each feature to weigh evenly.

Mars has the second-highest ESI in the solar system, clocking in at a value of 0.70. Venus, often called Earth's "twin," comes in at a measly

0.44. Venus's low ESI is due to the high solar flux (irradiation from the sun) and the impact of the high levels of carbon dioxide on the planet, which led to its having a runaway greenhouse effect; it is likely even too hot for most life-forms to exist, at least relative to the profiling complete by the twenty-first century. In planetary-science terms, this means that Venus is beyond its Komabayasi-Ingersoll limit, defined as the maximum solar flux a planet can handle without a runaway green-house effect. Once the planet gets too hot, its ESI decreases and is less desirable.

It is worth noting that the ESI, by its very nature, is biased toward Earth. Other researchers, including Madhu Jagadeesh, have defined Mars Similarity Indices (MSIs), and posit that the search for life and potentially habitable planets could also have MSI as a secondary metric (especially if Mars once had life). This metric could readily be made for any planet or moon as long as we have enough information on it—such as a Venus Similarity Index or a Titan Similarity Index. However, given that all understanding of life is based on what has developed on Earth, the best chance for a generation ship's mission to succeed would be through the use of the ESI and PHI metrics. Once we validate our plans to inhabit other worlds and create self-sustaining societies, then these planets' initial characteristics could be used to widen the net of possible worlds to which we could travel. These metrics will essentially allow us to "swipe left" on a planet that is not quite right, or "swipe right" on a planet that matches or piques our curiosity. Of course, additional probes and data would be required before starting anything long term.

OUR PLANETARY MATE(S)

So, then, where should we send the newly engineered humans and other organisms? Fortunately, as described earlier, we have rapidly pro-gressed from finding no Earthlike planets, in 2014, to finding thou-sands, with far more lying in wait for our discovery by 2500. Perhaps most strikingly, almost all the high-ESI and high-PHI planets discovered so far have been found in one very narrow and short view of the Milky Way (figure 11.4a). Far more planets will be found, some of which will likely even be closer than the majority of planets currently found, making the generation ship's travel time even more palatable.

Though these new findings may not even be needed, as many exoplanets cataloged in the twenty-first century could serve as potential human homes. The aptly named website Planetary Habitability Laboratory (PHL), kept at the University of Puerto Rico, tracks each new discovery that is either Subterran (Mars-size), Terran (Earth-size), or Superterran (Super-Earth/Mini-Neptunes). For many of the planets, there are only guesses at the actual surface temperature or atmosphere, making the PHI hard to measure, but at least the ESI has already given us some great candidates. Based on PHL's calculations at the time of publication, there are seventeen candidate planets greater than or equal to 0.8 ESI, and forty-two greater than or equal to 0.7.

A great candidate system is the seven-planet, TRAPPIST-1 system (planets a, b, c, d, e, f, g, and h). This system was analyzed through a collaboration between the scientists working on the Hubble, Kepler, and Spitzer space telescopes and the European Southern Observatory's SPECULOOS (Search for habitable Planets Eclipsing Ultra-cOOl Stars) telescope. Data collected, refined, and finally published in 2018 showed that the seven planets have masses of 0.3–1.2-fold that of Earth's, with similar densities, indicating their gravity would be tolerable, and even useful, for twenty-first-century humans. Some planets (c and e) seem to be mostly rocky, while others (b, d, f, g, and h) likely have some kind of watery, icy, or atmospheric exterior similar to Earth's. Excitingly, there is evidence that Planet d has a liquid-water ocean that makes up about 5 percent of its mass (Earth's water content only accounts for <0.1 percent of its mass). Further, most of the planets have evidence of a possible iron core, which bodes well for their magnetosphere. Our own ESI calculations on the TRAPPIST system and other close (relatively) exoplanets which could be potential homes for future life are visualized in figure 11.4b.

Though the TRAPPIST-1 system may be the best candidate if a generation ship were launched in the twenty-first century, it is not perfect. From the data collected thus far, all seven planets appear to be "tidally locked," just like Earth's moon and Titan with Saturn, where one side

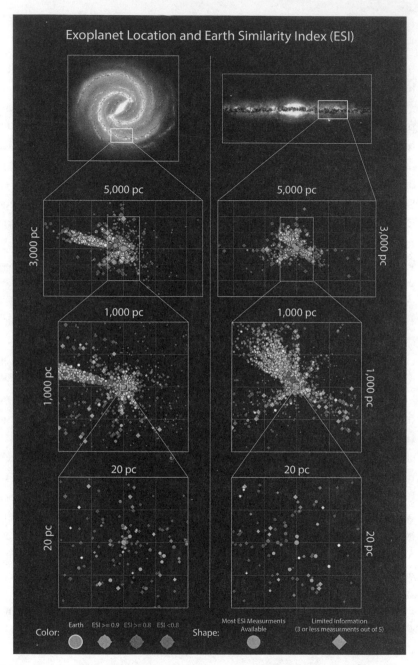

11.4a The location and similarity of all identified exoplanets: Most putative exoplanets that could be used for settlement by generation ships are within dozens or hundreds of parsecs (3.26 light-years) of Earth, either when examined from the top of the Milky Way (left) or the side-view of the Milky Way (right). The ESI for a variety of planets can be calculated based on the metrics of equilibrium temperature, density, solar flux, radius, and escape velocity. The highest-quality candidates are ≥0.9 (gray), secondary candidates are <0.9 but ≥0.8, and the lower-quality candidates are <0.8 (white). Exoplanets with only 3 or less metrics for ESI calculation are diamonds; those with 4 or all 5 metrics are circles. Data from https://exoplanetarchive.ipac.caltech.edu/. (See color plate 14.)

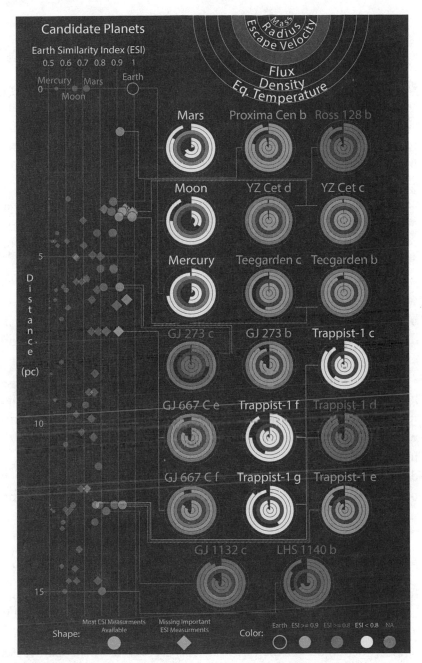

11.4b Best extracellular worlds based on location and similarity: ESI for candidate planets is shown relative along with their distance to Earth (y-axis). The ESI for a variety of planets can be calculated based on the metrics of equilibrium temperature, density, solar flux, radius, and escape velocity. The highest-quality candidates are ≥0.9 (gray), secondary candidates are <0.9 but ≥0.8, and the lower-quality candidates are <0.8 (white). Relative values are displayed as filled-in rows within the circular plots; light gray references missing data. (See color plate 15.)

of the planet is permanently facing its star. This means that these planets probably have very drastic and sharp differences in temperature between their permanently lit "light sides" and their ever-night "dark sides," which may also create unwanted stormy weather. The twilight regions which exist between these extremes, called the "terminator line," may therefore be the best regions to land and set up a new home. The Earth's terminator line is constantly in motion, appearing as sunset and sunrise, but the fixed TRAPPIST-1 planets' terminator lines may actually be an advantage. Crews could keep exploring in the "hotter" or "colder" parts of a planet, until they find the Goldilocks zone and a site that seems stable.

Our data are still very limited, both in the percentage of space that we have observed, and in the granularity of what we know about exoplanets we have found. Missions such as the "Project Starshot" will drastically aid in our understanding of what nearby solar systems and exoplanets actually look like by enabling us to analyze them closer up. Astronomers of the twenty-first century are much like an ophthalmologist trying to examine their patient's eyes from across the street amid traffic. We will know much more once we get closer. These missions will produce more reliable information on exoplanets and systems of interest—including data on surface chemistry, planetary activity, and data for PHI calculations. This information will be crucial to have well before the launch of any generation ship to maximize their success and minimize their need to go to plan B, C, D, or move to a new system.

PLANETARY ENGINEERING

In an ideal scenario, the generation ship would find an exoplanet that is immediately habitable. However, as with most of biology, survival is a question of access to resources, and there are many, many planetary or system conditions which can put this accessibility in jeopardy. If a planet is too hot, it might be impossible to survive at all. The extreme cold of a planet might be manageable—but even this would have limits on practicality and feasibility. The challenges of generation ship's mission will be compounded by the distance, bringing humans farther away from where they grew up, and the uncertainty which accompanies trying to analyze worlds that are full of unknown variables. The majority of how we plan this mission, what we bring with us, and how we set up our first exoplanet will be based on the success and challenges from inhabiting Mars, Titan, and even remote areas of Earth.

All of the technologies needed for mapping, monitoring, and adjusting Earth's atmosphere will be the basis for adjusting the atmosphere of new worlds. Even long-forgotten processes that once plagued Earth could be brought back to help. For example, chlorofluorocarbons (CFCs) are incredible greenhouse gases capable of making a new world hotter. This idea was first proposed in *The Greening of Mars* (by Michael Allaby and James Lovelock) as a way to heat Mars. Even though CFCs once destroyed the protective ozone in Earth's upper atmosphere, their future application in new worlds may provide us with the exact heat retention we need to survive, like a large blanket in the winter. The generation ship could even manufacture them, further improving its self-reliance and ability to have multiple backup plans.

If the planet is not yet habitable, next up will be engineering its atmosphere. This process could take hundreds, thousands, or tens of thousands of years, resulting in the generation ship being in orbit for many more generations beyond arrival. However, some small outposts would likely be built while waiting for complete touch down and further used for testing new chemistry and atmospheric engineering. An important step for the success of any large-scale engineering challenge, whether genetic or planetary, is to catalog and understand the accessibility, modularity, and interactions of all of its components. Ideal technologies would be ones capable of addressing multiple limitations of the world with few ingredients.

First, small biomes could be sent down to the surface. These could be used to help test new combinations of chemistry, as well as for atmospheric engineering. Fortunately, plants from Earth are an ideal model for planetary technology development. But how can you convert their consumption of sunlight and CO_2 to produce required reagents, such as energy and O_2, for the use on other worlds? This has been a long-standing question for the US Department of Energy, which now supports three Clean Energy Hubs (Nuclear Energy Modeling and Simulation, Energy Storage Research, and Critical Materials Institute), as well as the Joint Center for Artificial Photosynthesis. These initiatives have propelled the work of many scientists, including Harvard professors Pam Silver and Daniel G. Nocera, who first made a "bionic leaf" in 2011, capable of performing five to ten times better than non-bionic plants. Their "leaf" is a simple wafer of silicon and solid substrates that, when exposed to sunlight and water, split the water into hydrogen and oxygen.

Deployment of such bionic leaves in new worlds would enable quick fuel and air production, by creating hydrogen and oxygen simultaneously. However, these will rely on the availability of both light and water. While water is scarce on Mars, the red planet does have water in places as shallow as 2.5 cm (about one inch), based on a water "treasure map" published by NASA in 2019. Using such scanning technology would likely constrain the selection of exoplanets to those with accessible water, though that will likely be the case anyway since (Earth's) life seems to enjoy it.

Even though one of the primary jobs of plants on Earth is to photosynthesize, they actually aren't that good at it (about 1 percent efficient). Similar to the improvement seen with synthetic leaves, other biological life has been shown to actually do this better. Microalgae grown in bioreactors can sometimes reach 3 percent efficiency; and a maximum of 5–7 percent efficiency has been observed for microalgae in very specific conditions (called "bubble bioreactors"). Professor Nocera and other biologists, including Pam Silver, wanted to see just how far they could push this. Basing their work on that of Anthony Sinskey (at MIT), they took the unique bacterium *Raistonia eutropha*, which is capable of consuming hydrogen and CO_2 to produce ATP, and inserted new genes, which enabled it to further convert this ATP alcohol and a cobalt-phosphorus water-splitting catalyst, and even excrete its products in aerobic conditions.

This artificial photosynthesis system was far more efficient than that of natural photosynthesis and reached an impressive 10 percent. Beyond this advance, they even produced petrochemicals—including isopropanol, isobutanol, and isopentanol—which could be used in engines of the twenty-first century. This 2016 work was published to much fanfare because their system uses CO_2 in a carbon-neutral manner. Although it does not serve as a carbon sink to help with the excess carbon, it can help with the entire petrochemical industry that is "burning carbon to find more carbon to burn," continually exacerbating the greenhouse effect of CO_2 on Earth. On future planets, the degree of efficiency and types and proportion of gases released could be altered to help accelerate the changes of an atmosphere or a settlement into one that is more amenable to the crew.

Similarly, the NASA project Ecopoiesis Test Bed aims to use this exact idea of kick-starting a planetary engineering plan (terraforming) with "ecopoiesis" (from the Greek for "house" and "production") on Mars and eventually other worlds. The Ecopoiesis Test Bed project proposes to land at a site close to liquid water, completely seal itself off from the rest of the planet (for planetary protection), and then release extremophiles capable of surviving in their new home while sensing the presence or absence of metabolic products. Data generated from this project could then be captured by an orbiting satellite, and reports would aid in improving the chemical and biological composition of future probes and life. Over time, this will enable the engineering of strong and adaptive pioneering organisms capable of building entire ecosystems in the harshest of environments.

The biggest challenge in deep space exploration, especially generation ships where the destination may not be guaranteed, is the renewal of required resources. The farther we get from the sun, the more it will become just like any other star in the sky, and the less we will receive its nurturing light. Even if we have the ability to deploy these continually learning organism seedings on a planet, we would still need the materials required to synthesize the organisms and all other materials required to stay in orbit for how ever long such a project would take. As previously stated, the recycling and reuse ability on these ships will need to be extremely well designed, but the synthesis of materials will have to come from somewhere.

One solution to this would be extremely large ships, capable of storing the required raw building blocks the rest of the system feeds off of— which, of course, would still mean these cannot last indefinitely and would be much more challenging to build. Another solution, however, would be detailed planning on scavenging whatever is in interstellar space. This could be in the form of harvesting and actually using the constant barrage of radiation for the better or, likely, mining materials from asteroids and other space debris. The overall mission could be planned to maximize these encounters and restock materials whenever possible. If this plan were built into the generation ship missions, then the crew would not be despondent about a delay; rather, they would view it as part of the mission and their duty.

INTERSTELLAR, DIRECTED EVOLUTION

Over time, evolutionary changes will inevitably occur. Some rapid selection pressures (like with the silver fox) could quickly distinguish new humans or Guardian species from others, which itself would depend on what planet or ship their ancestors were from. Once this emerges, it will represent the first interstellar tracking system of life. Through molecular profiling and DNA sequencing of these changes, we will develop a catalog of how life changes around various stars and build a vast catalog of the panoply of life's adaptations. This vast genetic library can be continually compared to what we know about evolution on Earth, as well as to anywhere else humans and other life-forms will live, such as the moon, Mars, Titan, orbiters around different planets, and any world outside of our solar system.

Once identified, these specific molecular changes could further be analyzed, characterized, and examined for the ability to move into new biological systems. Those shown to be beneficial to living in specific locations (e.g., a moon in the TRAPPIST-1 system) could then be preemptively engineered into people before they visit (as discussed above and in figure 11.3). This would start a positive-feedback loop of undirected *and* directed evolution, which spans multiple worlds, stars, and eventually even galaxies. Eventually, these world-specific, engineered protocols could be preemptively engineered into the first settlers of new worlds by using a world similarity index (WSI), similar to ESI but without bias, to find what currently cataloged world it most resembles, further increasing mission success. Then, by extension, there would be a solar system similarity index (SSSI) and eventually galactic similarity index (GSI) for the best match for known life, and we could exquisitely discern those life-forms that are specific to any area of the mapped universe.

However, technology can be a double-edged sword. For any good a new discovery can bring, it could also be twisted in the wrong hands for a malicious purpose. As an example, imagine we find a specific type of nucleotide, or group of biological substrates, which is uniquely associated with living on a different world. This could inspire learning and continual improvements on directed evolution to ensure people can thrive under different conditions, but it can also lead to nefarious events whereby a "planetary terrorist" may engineer a virus that specifically attacks humans who have evolved to only live on this world in a specific chemical context. While the logistical and economic hurdles to such an effort are vast, they are not insurmountable, and history has shown that marginalized and suppressed groups, or even just those with different ideologies, can resort to drastic measures to fight for what they think is right. This potential biological warfare would need to be carefully monitored and safeguarded to ensure the safety of each world and its inhabitants.

Nonetheless, by 2401, we will have several generations of directed evolution, spanning human, fungal, bacterial, and planetary adaptations across multiple systems, orbiters, and ships. Combining this information with characteristics of known planets and moons will drive the positive-feedback loop between undirected and directed evolution. Given enough time, we will go beyond the planets and moons in our own solar system, exploring other stars, and someday even exploring other galaxies. We could potentially even engineer organisms capable of surviving space, not just persisting but even thriving in the void, including some being able to migrate between planets, as easily as flying within Earth's atmosphere. We might even have migrating species, such as tardigrades with solar wings, that move between newly settled planets like interstellar monarchs. The engineered life would be a new kind of poetry that moves between stars.

12

PHASE 10: OPTIMISTIC UNKNOWNS (BEYOND 2500)

A NEW HORIZON

Given the tools developed in the twenty-first through the twenty-sixth centuries, a world of undirected, unguided, and brutal evolution will no longer be inevitable. Humans will have a mastered an ability that truly sets us apart from other species —the ability to direct our own, and other organisms', evolution. With an ever-increasing quantification and awareness of extinction, we will continually improve our capacities as a Guardian species. With this new role comes many new opportunities and new duties (deontogenic), including the unique and first chance for a Guardian species to avoid extinction for ourselves and all other species. There is a chance that we will find bacterial life on other planets in our solar system, or even other intelligent life, by 2500. We could keep going onward, and outward, looking for new life. One big question is: For how long?

CLOSING THE DOOR OF OUR FIRST HOME

The end of the Earth will start with small changes. First, in about 50,000 years, the length of the day used for astronomical timekeeping will get longer, jumping to 86,401 seconds because of the moon's "pulling" on Earth and decelerating its rotation. Within a few million years, most of the constellations in the sky from the twenty-first century will be unrecognizable, and Cupid and Belinda (two moons of Uranus) will probably have collided, creating debris and a new source of ring material for the host planet.

In 100 million years, the sun's luminosity will increase by 1 percent, and things will really begin to heat up, but at levels to which humans could adapt. In about 180 million years, the Earth's rotation will have slowed enough that a day will become twenty-five instead of twenty-four hours long. In 250 million years, the northern coast of California may bump into Alaska, and in around 500 million years, a new supercontinent could appear on the surface of Earth. But then, around 600 million years from now, the moon will have moved far enough away from Earth that solar eclipses will no longer occur. The moon will still be revolving around Earth, but only partial eclipses will be seen.

In 700 million years, increasing luminosity of the sun will accelerate the weathering of surface rocks and trap more carbon dioxide as carbonate. As water evaporates from Earth's surface, plate tectonics will eventually slow and then stop, leading to the end of most volcanic activity. Without a way to recycle carbon into Earth's atmosphere, CO_2 levels will begin to fall, which will kill plants that use C3 photosynthesis (99 percent of known plant species). This will leave only the C4 plants (such as maize), which can function with less water and nutrients to complete photosynthesis.

For all these challenges, one (conceptually) simple solution would be to redirect asteroids or other planetary bodies toward Earth and adjust our orbit as the sun continues to enlarge. This would preserve the relative comfort of the planet we know today, but it would require energies and methods that do not yet exist. Attaching a giant engine to Earth and pushing it is not yet possible, even though it has been proposed, at least, in film (*Wandering Earth*). With so much plant life eliminated, oxygen levels will then begin to fall in the atmosphere as well, which would likely eliminate the ozone layer that protects most life from UV radiation.

By 800 million years, CO_2 levels will drop to where even the C4 photosynthesis-based plants could no longer survive. Without plant life and a way to cycle the carbon and oxygen in the atmosphere, most multicellular life will likely cease to exist. In a seminal book about this idea (*The Life and Death of Planet Earth*), Peter Ward and Donald Brownlee projected that perhaps some animal life could survive in the oceans, but even this is expected to be very difficult. Most of life is unlikely to be able to survive past 800 million years.

LEAVING THE INNER PLANETS

Then, in about a billion years, things get really hot; our sunshine will change even more dramatically. The sun will become about 10 percent brighter as a result of the proportion of heavier atoms used in fusion reactions, which will drive an increase in heat as well. The oceans, air, and planet will absorb this, and it could trigger a greenhouse effect that could make the future Earth resemble the hellscape of Venus. For anyone concerned about climate change today, the most damning number is one billion years, because that is when we need all possible technology to engineer our own planet not to overheat and kill everything we have ever known.

The challenges only get worse over time. Eventually (in around 3 billion years), the sun will increase in brightness even more (35 percent), which will then be enough energy to cause the oceans to boil, the ice caps to melt, and significant amounts of water vapor in the upper and middle atmospheres to float out into space. Life in this stage would be difficult to maintain on the surface of Earth, although Mars will suddenly be more temperate. Also, around this time, the magnetic fields protecting Earth will likely disappear, as the constant shifting of iron inside Earth's core (dynamo) stops.

Our own sun will begin to turn into a red giant in about 5 billion years, although some estimates suggest 5.5 billion years. Regardless of when it happens, at that time, anything left on Earth will be charred to a cinder and mostly reduced to atoms. Earth's orbit around the sun will drift outward, but even with the expanded radius, Earth will likely be very, very close to the outer radius of the new red giant sun (a scorcher). At that time, any of Earth's remaining atmosphere will evaporate into space, and it will be reduced to a lava sphere with floating "icebergs" of dense iron, drifting along in a 2,400K (2,130°C) bath.

One unexpected bonus of Earth's oblivion is that, in that late stage of the sun, effectively a new solar system will appear. The cold climate of Titan will suddenly become more balmy and akin to Martian conditions in the twenty-first century (around 200K). This means that even if we have not solved the crisis and gotten out of the solar system in 5 billion years, we could have another few hundred million or even a billion years to work things out in the outer reaches of the solar system. But even in the outer reaches, we cannot last forever.

In about 7 billion years, Mars and Earth will become tidally locked to the sun, with one side of each planet permanently facing the light. But the sun will keep enlarging, reaching its maximum size around 8 billion years from now, and almost certainly engulf the inner planets (Mercury, Venus, Earth, and likely Mars). Shortly thereafter, the sun will fizzle out and become a white dwarf with about 54 percent of its current mass and a much lower luminosity. This would be the sun's last big activity, and if any humans have not made it out of the solar system on a generation ship by then, they will have to be living on the outer planets (Saturn and Jupiter), their moons, or have found a way to live in a white dwarf system.

LOOKING FAR AHEAD

Estimates about the end of the universe are vague. The timing for the death of our universe depends on many things we cannot measure well, such as dark matter and dark energy, and the events are at least hundreds of billions of years away. Even the ability to ask such a daunting question was the purview of mythology and religion until Einstein's general-relativity equations in 1907 gave several possible solutions. However, each solution gives a different ending to the universe. What we do know is that the entire universe trends toward greater entropy, or disorder, at both the molecular and the planetary levels. Gravity and life are the only two forces of the universe that continually combat this entropy.

Our demise at the universe's end will likely occur in one of several ways, with two being the most likely. First, it could be from the universe's expanding endlessly, called the "big freeze" of the universe.

Planets continue to drift apart, then cells, molecules, atoms, and eventually even electrons and subatomic particles themselves will be too far away from each other to interact. To survive, life would need to arrange a perfectly sealed environment that controls its local gravity and that could recapture all the energy of the system, letting no energy escape. This would make it possible to persist in perpetuity, direct our own evolution (as described in previous chapters), and begin expeditions to reengineer the universe itself.

The other avenue of our demise could be in the "big crunch," where the dark matter and visible matter of the universe have enough density to drive all mass to continuously move closer together. As more and more matter condenses and coalesces toward a singular point, life would be directly bombarded by new suns and planets and continually fighting back against myriad celestial bodies that would impact or damage any inhabited planet. Here, too, the only path to survival is to reengineer the universe itself, whereby the structure of visible and dark matter would need to be either modified or converted into a discrete, contained entity. As we have seen throughout this book, deontogenic ethics requires us to modify the universe in order to preserve the universe.

Yet, it is possible that what preceded our own Big Bang was a previous incarnation of life, and that life could arise again in the new universe after this one. Moreover, life in the new universe could arise in a better, more stable, and more ideal form, with less war, famine, and disease. If we stopped this from happening, we would be doing harm to life's prospects, thus actually *violating* the deontogenic ethic. How could we know the impact of our decisions across multiple, consecutive universes?

UNLIMITED POTENTIAL

While we do not know the answers to all these questions, there are some things we do know. Earth and its resources will not last forever, many other worlds exist that we may be able to venture to, and spreading out the only known intelligent life—instead of keeping it all on the same vulnerable rock—is the best strategy to answer these questions, as well as coming up with potential resolutions to the fate of the universe.

Just as 1,000 years ago it would have been entirely inconceivable to extract specific cells from a body, engineer them with new genes that enable new abilities, and then reimplant them to cure a human and extend their life decades beyond what would otherwise be possible, the abilities of our progeny thousands, millions, or billions of years from now may be as unimaginable as manipulating dimensions themselves. However, the only way for our offspring, or any species, to have this possibility is if we ensure our (and their) survival.

The ideas in this book are not simply to prolong the survival of humans, but to achieve true cellular, molecular, and planetary liberty. This dream is not only for humans millions of years in the future, but for all species, from any kind of matter. Through the careful study of life around us and how genetic engineering may alter a cell or organism, we can eradicate many genetic diseases. We will further be able to develop technologies for people to choose what and who they want to be—and for their dreams, feelings, and passions to be further engrained into their own DNA within every cell in their body. An individual will no longer be (even partially) destined by what they inherited.

Humanity's curiosity will continue to develop like never before. As we continue to expand into the universe, many unimaginable discoveries await. For example, only in 2020 did we discover the aptly named "super-puff" or "cotton candy" planets that have roughly the density of the carnival treat yet are often as large as Jupiter.

The reality is, we don't know where life exists in the universe, how much there is, or frankly, even if there is life anywhere else. Indeed, one of the main limitations in the Drake equation, the Seager equation, and any other estimates of this type is that we are forced to have these discussions relative to Earth's life. The true landscape of "lifesignatures," as opposed to biosignatures that are used to examine planets for life, could span an entire catalog of different possibilities, and also different kinds of matter (e.g., dark matter life-forms or mechanical intelligences). While studying extremophiles and identifying the minimum number of genes with which a bacteria can survive are incredibly important to understand the potential limits to which specimens of Earth's life could exist on other planets, it is still likely a small window of the total potential of the universe's life. Much like human eyes' ability (at least in the twenty-first century) to see only an extremely small fraction of the electromagnetic spectrum (visible light), blind to all other wavelengths, we currently only see a very small fraction of the potential of life, compared to all life that may exist in the universe. This is quite literally a continually expanding universe, and many lifesignatures may be found.

How do we solve for the lifesignatures and the universe's entire possibilities of life? To do this we absolutely need at least two things: time and means to explore. It will take many years just to be able to start to live safely beyond Earth. It will then take longer to build necessary generation ships, and then even longer to travel. We need time to plan, test, and apply our findings to enable us to explore. Without exploring, we will never see what is possible off of Earth. This requires a shared vision that is beyond the scope of all of our lives, to which we all contribute.

Our understanding and definition of life has changed over time, even just on Earth. Before we understood cells and development, we used to think that a human fetus was essentially just a small baby: as if part of a set of Russian stacking dolls, the exact baby that would eventually emerge just simply got slightly larger and larger. Even the definition of what constitutes life is debated—from viruses to when a fetus is "living." Our fundamental definition and understanding of life will likely change as we explore many different solar systems, and these findings will likely spark an entirely new wave of life's development— much more than human development—unlike any other revolutionary period that predates it.

THE FINAL DUTY

Humans are capable of creating new universes in their heads through the selective firing of billions of neurons—currently existing only in their imagination, but, potentially, one day these creations may come to fruition. Preserving this adaptive, creative genesis of ideas, dreams, and goals is our duty, not only to our species and to this planet, but to all species and to all planets. Engineering life, by careful study and delineating the mechanistic details that enable life in the first place, will be the means of its own survival.

Yet the most important components of life's systems are the least reducible. The rise of complex traits and emergent properties which expand beyond an individual, such as philosophy, poetry, music, and other arts, as well as science and empiricism (and criticisms thereof), are challenging to embed into a "point-to-point" biological framework. Life is infinitely more complex than any single theory about it. Life continues to create new theories about itself and the universe. These theories can only be explored, and new dreams made, as long as consciousness is preserved. Evolution selects for the existing processes that are best suited for the current situation. Evolution does not have the ability to foresee what the next challenge will be, and evolution certainly does not know about its own impending doom.

But we do. We alone can act as the conscious, careful, and thoughtful incarnation of evolution, as Guardians, and direct it to prepare for long-term survival. All of the lessons above and genetic substrates for human ingenuity give a continually enriched, ever-fertile ground for adaptation and engineering, and only we possess this skill and the awareness of its fragility.

We should enable ourselves, or any subsequent sentience, to fulfill the final deontogenic duty, and to help the substrates of life to continue as long as possible—regardless of the type of life, the kind of matter from which it is built, the home planet, or the age of the universe.

If the universe itself, in all its limitless splendor, perpetual mystery, and continual change, were anthropomorphized and asked, "What kind of universe would you want to be?"

Mulling over the last of the subatomic particles, variegated energy, and sketches of plans, it would likely respond with its own deontogenic ethic and goal:

"A universe that creates new universes."

ACKNOWLEDGMENTS

The first seeds for this book were planted on my fifteenth birthday, when I was given a novel by my Aunt Annie and Uncle Jeff, *Foundation* by Isaac Asimov. That novel imagined the ease with which humans could live across an entire galaxy and how careful, long term planning could remove the suffering of humans and all other species. The idea never left my head, and I thank my aunt and uncle for the book, as well as the vision of Asimov.

Later, so many other people made it possible for me to get time to write and supported me along the way; I owe them unending gratitude. First, my wonderful wife and daughter, Joan Moriarty and Madeleine Mason-Moriarty, who supported me the whole time from book inception to completion (and through many 4:00 a.m. alarms). Joan is the one who would make sure the workers in these future worlds have all the rights they deserve. Joan carried the family while I worked on the book. Maddie always brought ice cream and smiles to me while writing, plus kisses on my head. The worlds I see in this book are ones I hope she sees and enjoys. Plus many others.

My amazing parents, Cory Mason and Roseann Mason, who always encouraged me to dream big, and in whose living room I have written hundreds of pages and who painstakingly did the index! My siblings, the awesome Rosemary Mason, who put me up in her basement for

weeks on end, and Cory Mason and Rebecca Mason, who always supported my spacecraft landing-strip ideas from the Model Organization
of American States and Model United Nations meetings in high school
and always encouraged dreams by the shores of Lake Michigan.

All past and current members of the Mason Lab at Weill Cornell
Medicine. The ever-guiding Chair of Physiology and Biophysics, Dr.
Harel Weinstein, and Drs. Laurie Glimcher and Augustine Choi, who
had endless faith in me as deans. Drs. Matthew State, Murat Gunel, and
Rick Lifton, the amazing "three tenors" who guided my postdoctoral
fellowship. Dr. Kevin White, who was my inspring thesis advisor at
Yale and now colleague at Tempus. Dr. Joel Dudley, a long-time friend
and cofounder of many companies, who was also a sounding board
for many of these ideas. Eric Lefkofsky for always being a visionary
and pushing bigger dreams. The superb Paul Jacobson, Bodi Zhang,
Laura Kunces, Sarah Pesce, Nathan Price, Lee Hood, John Earls, and
Nate Rickard at Onegevity/Aevum. The stellar "Biotians," Drs. Niamh
O'Hara and Dorottya Nagy-Szakal from Biotia. Dr. Yoav Gilad who
established that "the best is never good enough," and inspired me to
always work harder and smarter, and Dr. Zareen Gauhar, my own twin,
reminded me that Yoav was just "being Booboo" and reminded me of
the innate beauty in everything. Drs. George Church and Ting Wu,
with whom many evenings, dinners, and discussions at conferences
helped form ideas in this book.

Drs. Craig Kundrot, John Charles, Jenn Fogarty, Afshin Beheshti,
Marisa Covington, Mark Kelly, Scott Kelly, and Chuck Lloyd from
NASA, who always took time to grab lunch and think about these
ideas. Of course, the bravery and tenacity of Scott and Mark Kelly
cannot be overstated, and they have been wonderful partners in
the research that continues to this day. Also, the lab's awesome
team for the NASA Twins Study, especially Drs. Francine Garrett-
Bakelman, Cem Meydan, Matthew MacKay, Daniela Bezdan, and Ebrahim Afshinnekoo, and the other PIs and Co-PIs (Drs. Susan Bailey,
Mathias Basner, Andrew Feinberg, Stuart Lee, Emmanuel Mignot,
Brinda Rana, Scott Smith, Michael Snyder, and Fred Turek), many of
whom are part of ongoing follow-up work and plans for the next missions. Ari Melnick, who makes me love epigenetics more than I already

do. Also, Drs. Stacy Horner, Alexa McIntyre, and Nandan Gokhale, who worked over Christmas break to get the mitochondrial primers and MeRIP data completed and always inspire me to be a better scientist. Conan and Tammy Cecconie, Dan and Stacey Poulsen, in whose kitchens I wrote many pages early in the morning and who anchored many of the ethical discussions, along with Drs. Joe Martinelli, Scott Repa and also Scott Honsberger (IMSA). Drs. Max Lagally and Lloyd Smith from UW-Madison. Finally, thanks to Jack Balkin, Eddan Katz, Mike Seringhaus (carpet), and everyone at the Information Society Project (ISP) at Yale Law School for all their insightful ethical and legal discussions from 2005 to the current day.

Any good plan needs strong support. Igor Tulchinsky has supported this work and dream since we first met, including many fun discussions and generating ideas for the future, and I cannot thank him and WorldQuant enough for their support. Bill Ackman and Olivia Flatto from the Pershing Foundation have had endless faith in me and our lab, and I also cannot thank them enough.

I also need to thank some beats, including Deltron 3030, which is a great rap album set in the year 3030, with Kid Koala on the beats and superb lyrics by Dan the Automator and Del the Funky Homo Sapien. It was behind many of the keystrokes in this book, with a vision of agile rappers hopping between planets. Also, Martèn LeGrand's beautiful piano album, *Saskatchewan*, was the other engine of the keyboard's movements. And Logic's *No Pressure*. The final thanks go to Bob Moses for the album *Days Gone By*, with a wonderful blend of trip-hop and blues.

REFERENCES

CHAPTER 1

Garrett-Bakelman, Francine E., Manjula Darshi, Stefan J. Green, Ruben C. Gur, Ling Lin, Brandon R. Macias, et al. "The NASA Twins Study: A Multi-Omic, Molecular, Physiological, and Behavioral Analysis of a Year-Long Human Spaceflight." *Science* 364, no. 6436 (April 12, 2019).

Ingelsson, Björn, Daniel Söderberg, Tobias Strid, Anita Söderberg, Ann-Charlotte Bergh, Vesa Loitto, et al. "Lymphocytes Eject Interferogenic Mitochondrial DNA Webs in Response to CpG and Non-CpG Oligodeoxynucleotides of Class C." *PNAS* 115, no. 3 (2018): E478–E487.

Patti, Giuseppe, Andrea D'Ambrosio, Simona Mega, Gabriele Giorgi, Enrico Maria Zardi, Domenico Maria Zardi, et al. "Early Interleukin-1 Receptor Antagonist Elevation in Patients with Acute Myocardial Infarction." *Journal of American College of Cardiology* 43, no. 1 (2004): 35–38.

Schaefer, H. J., E. V. Benton, R. P. Henke, and J. J. Sullivan. "Nuclear Track Recordings of the Astronauts' Radiation Exposure on the First Lunar Landing Mission Apollo XI." *Radiation Research* 49, no. 2 (1972): 245–271.

CHAPTER 2

Kant, Immanuel. *Groundwork for the Metaphysics of Morals*. Riga, 1785.

Mill, John Stuart. *Utilitarianism*. London: Parker, Son, and Bourn, 1863.

Parfit, Derek. *Reasons and Persons*. Oxford: Clarendon Press, 1987.

Rawls, John. *A Theory of Justice*. Cambridge, MA: Belknap Press of Harvard University Press, 1971.

CHAPTER 3

Castro-Wallace, Sarah L., Charles Y. Chiu, Kristen K. John, Sarah E. Stahl, Kathleen H. Rubins, Alexa B. R. McIntyre, et al. "Nanopore DNA Sequencing and Genome Assembly on the International Space Station." *Scientific Data* 7, no. 1 (2017): 18022.

Cheng, Alexandre Pellan, Philip Burnham, John Richard Lee, Matthew Pellan Cheng, Manikkam Suthanthiran, Darshana Dadhania, and Iwijn De Vlaminck. "A Cell-Free DNA Metagenomic Sequencing Assay That Integrates the Host Injury Response to Infection." *PNAS* 116, no. 37 (2019): 18738–18744.

De Vlaminck, Iwijn, Hannah A. Valantine, Thomas M. Snyder, Calvin Strehl, Garrett Cohen, Helen Luikart, et al. "Circulating Cell-Free DNA Enables Noninvasive Diagnosis of Heart Transplant Rejection." *Science Translational Medicine* 6, no. 241 (2014): 241ra77.

ENCODE Project Consortium. "An Integrated Encyclopedia of DNA Elements in the Human Genome." *Nature* 489, no. 7414 (2012): 57–74.

Hood, Leroy. "Tackling the Microbiome." *Science* 336, no. 6086 (2012): 1209.

Karczewski, Konrad J., Laurent C. Francioli, Grace Tiao, Beryl B. Cummings, Jessica Alföldi, Qingbo Wang, et al. "Variation across 141,456 Human Exomes and Genomes Reveals the Spectrum of Loss-of-Function Intolerance across Human Protein-Coding Genes." *biorxiv*, January 30, 2019 (preprint). https://doi.org/10.1101/531210.

Lindqvist, D., J. Fernström, C. Grudet, L. Ljunggren, L. Träskman-Bendz, L. Ohlsson, and Å Westrin. "Increased Plasma Levels of Circulating Cell-Free Mitochondrial DNA in Suicide Attempters: Associations with HPA-Axis Hyperactivity." *Translational Psychiatry* 6 (2016): e971.

Maier, Lisa, Mihaela Pruteanu, Michael Kuhn, Georg Zeller, Anja Telzerow, Exene Erin Anderson, et al. "Extensive Impact of Non-antibiotic Drugs on Human Gut Bacteria." *Nature* 555, no. 7698 (2018): 623–628.

McIntyre, Alexa B. R., Noah Alexander, Kirill Grigorev, Daniela Bezdan, Heike Sichtig, Charles Y. Chiu, and Christopher E. Mason. "Single-Molecule Sequencing Detection of N6-Methyladenine in Microbial Reference Materials." *Nature Communications* 10, no. 1 (2019): 579.

McIntyre, Alexa B. R., Lindsay Rizzardi, Angela M. Yu, Noah Alexander, Gail L. Rosen, Douglas J. Botkin, et al. "Nanopore Sequencing in Microgravity." *Nature Partner Journals (npj) Microgravity* 2 (2016): 16035.

MetaSUB International Consortium. "The Metagenomics and Metadesign of the Subways and Urban Biomes (MetaSUB) International Consortium Inaugural Meeting Report." *Microbiome* 4, no. 1 (2016): 24.

Miga, Karen H., Sergey Koren, Arang Rhie, Mitchell R. Vollger, Ariel Gershman, Andrey Bzikadze, et al. "Telomere-to-Telomere Assembly of a Complete Human X Chromosome." *biorxiv*, August 16, 2019 (preprint). https://doi.org/10.1101/735928.

CHAPTER 4

Barrangou, Rodolphe, Christophe Fremaux, Hélène Deveau, Melissa Richards, Patrick Boyaval, Sylvain Moineau, et al. "CRISPR Provides Acquired Resistance against Viruses in Prokaryotes." *Science* 315 (2007): 1709–1712.

Bernstein, L. "Epidemiology of Endocrine-Related Risk Factors for Breast Cancer." *Journal of Mammary Gland Biology and Neoplasia* 7, no. 1 (2002): 3–15.

Gootenberg, Jonathan S., Omar O. Abudayyeh, Jeong Wook Lee, Patrick Essletzbichler, Aaron J. Dy, Julia Joung, Vanessa Verdin, et al. "Nucleic Acid Detection with CRISPR-Cas13a/C2c2." *Science* 356, no. 6336 (2017): 438–442.

Hashimoto, Takuma, Daiki D. Horikawa, Yuki Saito, Hirokazu Kuwahara, Hiroko Kozuka-Hata, Tadasu Shin-I., et al. "Extremotolerant Tardigrade Genome and Improved Radiotolerance of Human Cultured Cells by Tardigrade-Unique Protein." *Nature Communications* 7 (2016): 12808.

Jinek, Martin, Krzysztof Chylinski, Ines Fonfara, Michael Hauer, Jennifer A. Doudna, and Emmanuelle Charpentier. "A Programmable Dual-RNA-Guided DNA Endonuclease in Adaptive Bacterial Immunity." *Science* 337 (2012): 816–821.

Lander, Eric S. "The Heroes of CRISPR." *Cell* 164, nos. 1–2 (2016): 18–28.

MacKay, Matthew, Ebrahim Afshinnekoo, Jonathan Rub, Ciaran Hassan, Mihir Khunte, Nithyashri Baskaran, et al. "The Therapeutic Landscape for Cells Engineered with Chimeric Antigen Receptors." *Nature Biotechnology* 6, no. 8 (2020): 120–128.

Sulak, Michael, Lindsey Fong, Katelyn Mika, Sravanthi Chigurupati, Lisa Yon, Nigel P. Mongan, et al. "TP53 Copy Number Expansion Is Associated with the Evolution of Increased Body Size and an Enhanced DNA Damage Response in Elephants." *eLife* 5 (2016): e11994.

CHAPTER 5

Chen, Guojun, Amr A. Abdeen, Yuyuan Wang, Pawan K. Shahi, Samantha Robertson, Ruosen Xie, et al. "A Biodegradable Nanocapsule Delivers a Cas9 Ribonucleoprotein Complex for In Vivo Genome Editing." *Nature Nanotechnology* 14, no. 10 (2019): 974–980.

Desai, Pinkal, Nuria Mcncia-Trinchant, Oleksandr Savenkov, Michael S. Simon, Gloria Cheang, Sangmin Lee, et al. "Somatic Mutations Precede Acute Myeloid Leukemia Years before Diagnosis." *Nature Medicine* 24 (2018): 1015–1023.

Unal Eroglu, Arife, Timothy S. Mulligan, Liyun Zhang, David T. White, Sumitra Sengupta, Cathy Nie, et al. "Multiplexed CRISPR/Cas9 Targeting of Genes Implicated in Retinal Regeneration and Degeneration." *Frontiers in Cell and Developmental Biology* 6 (2018): 88.

CHAPTER 6

Li, Zhi-Kun, Le-Yun Wang, Li-Bin Wang, Gui-Hai Feng, Xue-Wei Yuan, Chao Liu, et al. "Generation of Bimaternal and Bipaternal Mice from Hypomethylated Haploid ESCs with Imprinting Region Deletions." *Cell Stem Cell* 23, no. 5 (2018): 665–676.e4.

MacKay, Matthew, Ebrahim Afshinnekoo, Jonathan Rub, Ciaran Hassan, Mihir Khunte, Nithyashri Baskaran, et al. "The Therapeutic Landscape for Cells Engineered with Chimeric Antigen Receptors." *Nature Biotechnology* 6, no. 8 (2020): 120–128.

Posner, Rachel, Itai Antoine Toker, Olga Antonova, Ekaterina Star, Sarit Anava, Eran Azmon, et al. "Neuronal Small RNAs Control Behavior Transgenerationally." *Cell* 177, no. 7 (2019): 1814–1826.e15.

Vierbuchen, Thomas, Austin Ostermeier, Zhiping P. Pang, Yuko Kokubu, Thomas C. Südhof, and Marius Wernig. "Direct Conversion of Fibroblasts to Functional Neurons by Defined Factors." *Nature* 463, no. 7284 (2010): 1035–1041.

CHAPTER 7

Esvelt, Kevin. "When Are We Obligated to Edit Wild Creatures?" *LeapsMag*, August 30, 2019.

Gibson, Daniel G., John I. Glass, Carole Lartigue, Vladimir N. Noskov, Ray-Yuan Chuang, Mikkel A. Algire, et al. "Creation of a Bacterial Cell Controlled by a Chemically Synthesized Genome." *Science* 329, no. 5987 (2010): 52–56.

Hoshika, Shuichi, Nicole A. Leal, Myong-Jung Kim, Myong-Sang Kim, Nilesh B. Karalkar, Hyo-Joong Kim, et al. "Hachimoji DNA and RNA: A Genetic System with Eight Building Blocks." *Science* 363, no. 6429 (2019): 884–887.

Hutchison, Clyde A., Ray-Yuan Chuang, Vladimir N. Noskov, Nacyra Assad-Garcia, Thomas J. Deerinck, Mark H. Ellisma, et al. "Design and Synthesis of a Minimal Bacterial Genome." *Science* 351, no. 6280 (2016): aad6253.

Mohan, Malli, Ganesh Babu, Moogega Cooper Stricker, and Kasthuri Venkateswaran. "Microscopic Characterization of Biological and Inert Particles Associated with Spacecraft Assembly Cleanroom." *Scientific Reports* 9, no. 1 (2019): 14251.

Nagel, Thomas. "What Is It Like to Be a Bat?" *Philosophical Review* 83, no. 4 (Oct. 1974): 435–450.

Ostrov, Nili, Matthieu Landon, Marc Guell, Gleb Kuznetsov, Jun Teramoto, Natalie Cervantes, et al. "Design, Synthesis, and Testing toward a 57-Codon Genome." *Science* 353, no. 6301 (2016): 819–822.

Voorhies, Alexander A., C. Mark Ott, Satish Mehta, Duane L. Pierson, Brian E. Crucian, Alan Feiveson, et al. "Study of the Impact of Long-Duration Space Missions at the International Space Station on the Astronaut Microbiome." *Scientific Reports* 9 (2019): 9911.

Yue, Yanan, Yinan Kan, Weihong Xu, Hong-Ye Zhao, Yixuan Zhou, Xiaobin Song, et al. "Extensive Mammalian Germline Genome Engineering." *biorxiv* (preprint), December 17, 2019. https://www.biorxiv.org/content/10.1101/2019.12.17.876862v1 .full.pdf.

CHAPTER 8

Antiretroviral Therapy Cohort Collaboration. "Survival of HIV-Positive Patients Starting Antiretroviral Therapy between 1996 and 2013: A Collaborative Analysis of Cohort Studies." *Lancet HIV* 4, no. 8 (2017): e349–e356.

Christner, Brent C., John C. Priscu, Amanda M. Achberger, Carlo Barbante, Sasha P. Carter, Knut Christianson, et al., and the WISSARD Science Team. "A Microbial Ecosystem beneath the West Antarctic Ice Sheet." *Nature* 512, no. 7514 (2014): 310–313.

Li, Yi, Chang-Xin Shi, Karen L. Mossman, Jack Rosenfeld, Yong Chool Boo, and Herb E. Schellhorn. "Restoration of Vitamin C Synthesis in Transgenic $Gulo^{-/-}$ Mice by Helper-Dependent Adenovirus Based Expression of Gulonolactone Oxidase." *Human Gene Therapy* 19, no. 12 (2008): 1349–1358.

Pontin, Jason. "The Genetics (and Ethics) of Making Humans Fit for Mars." *Wired*, August 7, 2018.

Regalado, Antonio. "Engineering the Perfect Astronaut." *MIT Technology Review*, April 15, 2017.

CHAPTER 9

Blue, Rebecca S., Tina M. Bayuse, Vernie R. Daniels, Virginia E. Wotring, Rahul Suresh, Robert A. Mulcahy, and Erik L. Antonsen. "Supplying a Pharmacy for NASA Exploration Spaceflight: Challenges and Current Understanding." *npj Microgravity* 5 (2019): art. 14.

Goddard, Robert. "The Ultimate Migration." January 16, 1918.

Gros, Claudius. "Why Planetary and Exoplanetary Protection Differ: The Case of Long Duration Genesis Missions to Habitable but Sterile M-Dwarf Oxygen Planets." *Acta Astronautica*. 157 (2019): 263–267.

Ilgrande, Chiara, Tom Defoirdt, Siegfried E. Vlaeminck, Nico Boon, and Peter Clauwaert. "Media Optimization, Strain Compatibility, and Low-Shear Modeled Microgravity Exposure of Synthetic Microbial Communities for Urine Nitrification in Regenerative Life-Support Systems." *Astrobiology* 19 (2019): 1353–1362.

Jansen, Heiko T., Shawn Trojahn, Michael W. Saxton, Corey R. Quackenbush, Brandon D. Evans Hutzenbiler, O. Lynne Nelson, et. al. "Hibernation Induces Widespread Transcriptional Remodeling in Metabolic Tissues of the Grizzly Bear." *Communications Biology* 2 (2019): 336.

Lemmon, Zachary H., Nathan T. Reem, Justin Dalrymple, Sebastian Soyk, Kerry E. Swartwood, Daniel Rodriguez-Leal, et al. "Rapid Improvement of Domestication Traits in an Orphan Crop by Genome Editing." *Nature Plants* 4, no. 10 (2018): 766–770.

Nangle, S. N., M. Y. Wolfson, et al. "The Case for Biotechnology on Mars." *Nature Biotechnology* 3, no. 4 (2020): 401–407.

Parkin, Kevin. "The Breakthrough Starshot System Model." *Acta Astronautica* 152 (2018): 370–384.

Schrader, Jens, Martin Schilling, Dirk Holtmann, Dieter Sell, Murillo Villela Filho, Achim Marx, and Julia A. Vorholt. "Methanol-Based Industrial Biotechnology: Current Status and Future Perspectives of Methylotrophic Bacteria." *Trends in Biotechnology* 27 (2009): 107–115.

Ullah, Kifayat, Mushtaq Ahmad, Sofia, Vinod Kumar Sharma, Pengme Lu, Adam Harvey, et al. "Algal Biomass as a Global Source of Transport Fuels: Overview and Development Perspectives." *Progress in Natural Science: Materials International* 24 (2014): 329–339.

CHAPTER 10

Anglada-Escudé, Guillem, Pedro J. Amado, John Barnes, Zaira M. Berdiñas, R. Paul Butler, Gavin A. L. Coleman, et al. "A Terrestrial Planet Candidate in a Temperate Orbit around Proxima Centauri." *Nature* 536, no. 7617 (2016): 437–440.

Dugatkin, Lee Alan. "The Silver Fox Domestication Experiment." *Evolution: Education and Outreach* 11 (2018): 16.

Fraunhofer, Joseph. "Bestimmung des Brechungs- und des Farben-Zerstreuungs—Vermögens verschiedener Glasarten, in Bezug auf die Vervollkommnung achromatischer Fernröhre." [Determination of the refractive and color-dispersing power of different types of glass, in relation to the improvement of achromatic telescopes.] *Denkschriften der Königlichen Akademie der Wissenschaften zu München* [Memoirs of the Royal Academy of Sciences in Munich] 5 (1814–1815): 193–226.

Kukekova, Anna V., Jennifer L. Johnson, Xueyan Xiang, Shaohong Feng, Shiping Liu, Halie M. Rando, et al. "Red Fox Genome Assembly Identifies Genomic Regions

Associated with Tame and Aggressive Behaviours." *Nature Ecology Evolution* 2 (2018): 1479–1491.

Lissauer, J. "Three Planets for Upsilon Andromedae." *Nature* 398 (1999): 659.

Wang, Xu, Lenore Pipes, Lyudmila N. Trut, Yury Herbeck, Anastasiya V. Vladimirova, Rimma G. Gulevich, et al. "Genomic Responses to Selection for Tame/Aggressive Behaviors in the Silver Fox (*Vulpes vulpes*)." *PNAS* 115 (2018): 10398–10403.

Wohlforth, Charles, and Amanda R. Hendrix. *Beyond Earth: Our Path to a New Home in the Planets*. Pantheon, 2016.

Wolszczan, A., and D. Frail. "A Planetary System around the Millisecond Pulsar PSR1257 + 12." *Nature* 355 (1992): 145–147.

CHAPTER 11

Allaby, Michael, and James Lovelock. *The Greening of Mars*. New York: Warner Books, 1985.

Barnes, Elizabeth. *The Minority Body*. 2016. New York: Oxford University Press.

Baruch, S. "Preimplantation Genetic Diagnosis and Parental Preferences: Beyond Deadly Disease." *Houston Journal of Health Law Policy* (2008): 245–268.

Harrison, Steven M., Erin R. Riggs, Donna R. Maglott, Jennifer M. Lee, Danielle R. Azzariti, Annie Niehaus, et al. "Using ClinVar as a Resource to Support Variant Interpretation." *Current Protocols in Human Genetics* 89 (2016): 8.16.1–8.16.23.

Holtkamp, Kim C. A., Inge B. Mathijssen, Phillis Lakeman, Merel C. van Maarle, Wybo J. Dondorp, Lidewij Henneman, and Martina C. Cornel. "Factors for Successful Implementation of Population-Based Expanded Carrier Screening: Learning from Existing Initiatives." *European Journal of Public Health* 27, no. 2 (2017): 372–377.

Liu, Chong, Brendan C. Colón, Marika Ziesack, Pamela A. Silver, and Daniel G. Nocera. "Water Splitting-Biosynthetic System with CO_2 Reduction Efficiencies Exceeding Photosynthesis." *Science* 352, no. 6290 (2016): 1210–1213.

Schulze-Makuch, Dirk, Abel Méndez, Alberto G. Fairén, Philip von Paris, Carol Turse, Grayson Boyer, et al. "A Two-Tiered Approach to Assessing the Habitability of Exoplanets." *Astrobiology* (2011): 1041–1052.

CHAPTER 12

Brownlee, Donald E. "Planetary Habitability on Astronomical Time Scales." In *Heliophysics: Evolving Solar Activity and the Climates of Space and Earth*, edited by Carolus J. Schrijver and George L. Siscoe, 79–98. Cambridge: Cambridge University Press, 2010.

Caldeira, Ken, and James F. Kasting. "The Life Span of the Biosphere Revisited." *Nature* 360 (1992): 721–723.

Catling, David C., Joshua Krissansen-Totton, Nancy Y. Kiang, David Crisp, Tyler D. Robinson, Shiladitya DasSarma, et al. "Exoplanet Biosignatures: A Framework for Their Assessment." *Astrobiology* 18, no. 6 (2018): 709–738.

Krissansen-Totton, Joshua, Stephanie Olson, and David C. Catling. "Disequilibrium Biosignatures over Earth History and Implications for Detecting Exoplanet Life." *Science Advances* 4, no. 1 (2018): eaao5747.

Loeb, Abraham, Rafael A. Batista, and David Sloan. "Relative Likelihood for Life as a Function of Cosmic Time." *Journal of Cosmology and Astroparticle Physics* 8 (2016): 40.

Singer, Peter. *Animal Liberation*. New York: HarperCollins, 1975.

Waszek, Lauren, Jessica Irving, and Arwen Deuss. "Reconciling the Hemispherical Structure of Earth's Inner Core with Its Super-Rotation." *Nature Geoscience* 4, no. 4 (2011): 264–267.

INDEX